COST AND
IN PROJECTS

CUSTOM UPDATE EDITION

COST AND VALUE MANAGEMENT IN PROJECTS

CUSTOM UPDATE EDITION

Ray R. Venkataraman and Jeffrey K. Pinto

John Wiley & Sons, Inc.

This book is printed on acid-free paper. ∞

Copyright © 2012 by John Wiley & Sons, Inc. All rights reserved.

Published by John Wiley & Sons, Inc., Hoboken, New Jersey.
Published simultaneously in Canada.

No part of this publication may be reproduced, stored in a retrieval system, or transmitted in any form or by any means, electronic, mechanical, photocopying, recording, scanning, or otherwise, except as permitted under Section 107 or 108 of the 1976 United States Copyright Act, without either the prior written permission of the Publisher, or authorization through payment of the appropriate per-copy fee to the Copyright Clearance Center, Inc., 222 Rosewood Drive, Danvers, MA 01923, 978-750-8400, fax 978-646-8600, or on the web at www.copyright.com. Requests to the Publisher for permission should be addressed to the Permissions Department, John Wiley & Sons, Inc., 111 River Street, Hoboken, NJ 07030, (201) 748-6011, fax (201) 748-6008, or online at http://www.wiley.com/go/permissions.

Limit of Liability/Disclaimer of Warranty: While the publisher and author have used their best efforts in preparing this book, they make no representations or warranties with respect to the accuracy or completeness of the contents of this book and specifically disclaim any implied warranties of merchantability or fitness for a particular purpose. No warranty may be created or extended by sales representatives or written sales materials. The advice and strategies contained herein may not be suitable for your situation. You should consult with a professional where appropriate. Neither the publisher nor author shall be liable for any loss of profit or any other commercial damages, including but not limited to special, incidental, consequential, or other damages.

For general information on our other products and services please contact our Customer Care Department within the U.S. at 800-762-2974, outside the U.S. at 317-572-3993 or fax 317-572-4002.

Wiley also publishes its books in a variety of electronic formats. Some content that appears in print, however, may not be available in electronic books. For more information about Wiley products, visit our Web site at www.wiley.com.

Library of Congress Cataloging-in-Publication Data:

Venkataraman, Ray R.
　Cost and value management in projects / Ray R. Venkataraman and Jeffrey K. Pinto.
　　　p. cm.
　Includes bibliographical references and index.
　ISBN 978-1-118-27310-4
　1. Project management.　　2. Cost control.　　3. Value analysis (Cost control)
I. Pinto, Jeffrey K.　　II. Title.
　HD69.P75V46 2008
　658.4′04—dc22a

2007024563

Printed in the United States of America.

10 9 8 7 6 5 4

Contents

1 **Introduction to the Challenge of Cost and Value Management in Projects** — 1
   1.1 Importance of Cost and Value Management in Projects — 2
   1.2 Keys to Effective Project Cost Management — 6
   1.3 Essential Features of Project Value Management — 8
   1.4 Organization of the Book — 9
   References — 14
   Key Terms — 15

2 **Project Needs Assessment, Concept Development, and Planning** — 17
   2.1 Needs Identification — 19
   2.2 Conceptual Development — 22
   2.3 The Statement of Work — 24
   2.4 Project Planning — 27
   2.5 Project Scope Definition — 28
      2.5.1 Purpose of the Scope Definition Document — 28
      2.5.2 Elements of the Scope Definition Document — 28
      2.5.3 Project Scope Changes — 30
   2.6 Work Breakdown Structure — 32
      2.6.1 Types of Work Breakdown Structures — 33
      2.6.2 Work Breakdown Structure Development — 35
      2.6.3 Coding of Work Breakdown Structures — 38
      2.6.4 Integrating the WBS and the Organization — 38
      2.6.5 Guidelines for Developing a Work Breakdown Structure — 41
   References — 42
   Key Terms — 42

3 **Cost Estimation** — 43
   3.1 Importance of Cost Estimation — 44

3.2	Prolems of Cost Estimation		45
3.3	Sources and Categories of Project Costs		49
3.4	Cost Estimating Methods		51
3.5	Cost Estimation Process		56
	3.5.1	Creating the Detailed Estimate	56
3.6	Allowances for Contingencies in Cost Estimation		59
3.7	The Use of Learning Curves in Cost Estimation		61
References			64
Key Terms			65
Appendix			67

4 Project Budgeting — 83

- 4.1 Issues in Project Budgeting — 84
- 4.2 Developing a Project Budget — 85
  - 4.2.1 Issues in Creating a Project Budget — 85
- 4.3 Approaches to Developing a Project Budget — 85
  - 4.3.1 Top-down Budgeting — 86
    - Top-down Budgeting: Advantages — 86
    - Top-down Budgeting: Disadvantages — 87
  - 4.3.2 Bottom-up Budgeting — 88
    - Bottom-up Budgeting: Advantages — 89
    - Bottom-up Budgeting: Disadvantages — 90
- 4.4 Activity-based Costing — 90
  - 4.4.1 Steps in Activity-based Costing — 90
  - 4.4.2 Cost Drivers in Activity-based Costing — 91
  - 4.4.3 Sample Project Budget 1 — 91
  - 4.4.4 Sample Project Budget 2 — 92
- 4.5 Program Budgeting — 93
  - 4.5.1 Time-phased Budgets — 93
  - 4.5.2 Tracking Chart — 94
- 4.6 Developing a Project Contingency Budget — 95
  - 4.6.1 Allocation of Contingency Funds — 95
  - 4.6.2 Drawbacks of Contingency Funding — 96
  - 4.6.3 Advantages of Contingency Funding — 97
- 4.7 Issues in Budget Development — 98
- 4.8 Crashing the Project: Budget Effects — 99
    - Crashing Project Activities—Decision Making — 100
- References — 104
- Key Terms — 104

Contents

5 Project Cost Control — 105
- 5.1 Overview of the Project Evaluation and Control System — 105
  - 5.1.1 Project Control Process — 106
- 5.2 Integrating Cost and Time in Monitoring Project Performance: The S-Curve — 107
- 5.3 Earned Value Management — 111
- 5.4 Earned Value Management Model — 112
- 5.5 Fundamentals of Earned Value — 114
- 5.6 EVM Terminology — 114
- 5.7 Relevancy of Earned Value Management — 115
- 5.8 Conducting an Earned Value Analysis — 117
- 5.9 Performing an Earned Value Assessment — 119
- 5.10 Managing a Portfolio of Projects with Earned Value Management — 122
- 5.11 Important Issues in the Effective Use of Earned Value Management — 123
- References — 126
- Key Terms — 126

6 Cash Flow Management — 127
- 6.1 The Concept of Cash Flow — 127
- 6.2 Cash Flow and the Worth of Projects — 131
  - 6.2.1 The Time Value of Money, and Techniques for Determining It — 132
  - 6.2.2 Applying Discounting to Project Cash Flow — 134
- 6.3 Payment Arrangements — 137
  - 6.3.1 Cost-reimbursable Arrangements — 138
  - 6.3.2 Payment Plans — 140
  - 6.3.3 Claims and Variations — 142
  - 6.3.4 Cost Variation Due to Inflation and Exchange Rate Fluctuation — 144
  - 6.3.5 Price Incentives — 145
  - 6.3.6 Retentions — 146
- References — 148
- Key Terms — 148

7 Financial Management in Projects — 149
- 7.1 Financing of Projects Versus Project Finance — 150
- 7.2 Principles of Financing Projects — 150
- 7.3 Types and Sources of Finance — 151

	7.4	Sources of Finance	153
	7.5	Cost of Financing	153
	7.6	Project Finance	154
	7.7	The Process of Project Financial Management	156
		7.7.1 Conducting Feasibility Studies	156
		7.7.2 Planning the Project Finance	156
		7.7.3 Arranging the Financial Package	157
		7.7.4 Controlling the Financial Package	157
		7.7.5 Controlling Financial Risk	158
		7.7.6 Options Models	159
	References		161
	Key Terms		161
8	Value Management		**163**
	8.1	Concept of Value	163
	8.2	Dimensions and Measures of Value	166
	8.3	Overview of Value Management	167
		8.3.1 Definition	168
		8.3.2 Scope	168
		8.3.3 Key Principles of VM	168
		8.3.4 Key Attributes of VM	169
	8.4	Value Management Terms	169
	8.5	Need for Value Management in Projects	171
	8.6	The Value Management Approach	171
		8.6.1 Cross-functional Framework	172
		8.6.2 Use of Functions	172
		8.6.3 Structured Decision Process	172
	8.7	The VM Process	173
	8.8	Benefits of Value Management	175
	8.9	Other VM Requirements	175
	8.10	Value Management Reviews	176
	8.11	Relationship between Project Value and Risk	180
	8.12	Value Management as an Aid to Risk Assessment	181
	8.13	An Example of How VM and Risk Management Interrelate	182
	References		184
	Key Terms		184
9	Change Control and Configuration Management		**185**
	9.1	Causes of Changes	186
	9.2	Influence of Changes	190

Contents ix

9.3	Configuration Management	191
9.4	Configuration Management Standards	192
9.5	The CM Process	193
9.6	Control of Changes	196
9.7	Change Control Procedure and Configuration Control	197
9.8	Responsibility for the Control of Changes	200
9.9	Crisis Management	201
9.10	An Example of Configuration Management	202
	References	206
	Key Terms	207

10 Supply Chain Management 209

10.1	What Is Supply Chain Management?	210
10.2	The Need to Manage Supply Chains	211
10.3	SCM Benefits	212
10.4	Critical Areas of SCM	213
	10.4.1 Customers	213
	10.4.2 Suppliers	213
	10.4.3 Design and Operations	214
	10.4.4 Logistics	214
	10.4.5 Inventory	215
10.5	SCM Issues in Project Management	215
10.6	Value Drivers in Project Supply Chain Management	217
10.7	Optimizing Value in Project Supply Chains	220
	10.7.1 Total Quality Management	220
	10.7.2 Choosing the Right Supply Chain	221
10.8	Project Supply Chain Process Framework	221
	10.8.1 Procurement	221
	Supply Chain Relationships	223
	Supplier Development	224
	10.8.2 Conversion	224
	10.8.3 Delivery	225
10.9	Integrating the Supply Chain	225
10.10	Performance Metrics in Project Supply Chain Management	227
10.11	Project Supply Chain Metrics and the Supply Chain Operations Reference (SCOR) Model	230
10.12	Future Issues in Project Supply Chain Management	231
	References	232
	Key Terms	234

11 Quality Management in Projects — 235
- 11.1 Definition of Quality in Projects — 235
- 11.2 Elements of Project Quality — 237
  - 11.2.1 The Project's Product — 238
    - Quality Engineering — 239
  - 11.2.2 Management Processes — 243
  - 11.2.3 Quality Planning — 243
  - 11.2.4 Quality Assurance (QA) — 244
  - 11.2.5 Quality Control — 245
  - 11.2.6 Corporate Culture — 245
- 11.3 Total Quality Management (TQM) in Projects — 245
- 11.4 Quality Management Methods for a Project Organization — 247
  - 11.4.1 The Six Sigma Methodology — 249
  - 11.4.2 The Six Sigma Model for Projects — 250
  - 11.4.3 Application of Six Sigma in Software Project Management — 251
- 11.5 Quality Standards for Projects — 252
- References — 253
- Key Terms — 254

12 Integrating Cost and Value in Projects — 255
- 12.1 The Project Value Chain — 255
- 12.2 Project Value Chain Analysis — 257
- 12.3 Sources and Strategies for Integrating Cost and Value in Projects — 259
  - 12.3.1 The Project's Inbound Supply Chain — 260
  - 12.3.2 Project Design — 260
  - 12.3.3 Project Development — 265
  - 12.3.4 Project Delivery/Implementation — 267
    - Life-cycle Costing — 268
  - 12.3.5 Costs of Project Life Cycle Employing the LCC Model — 271
- 12.4 Integrated Value and Risk Management — 272
- 12.5 The Project Cost and Value Integration Process — 274
- References — 277
- Key Terms — 278

Index — 279

Chapter 1

Introduction to the Challenge of Cost and Value Management in Projects

LEARNING OBJECTIVES

- Identify the importance of cost and value management in projects.
- Describe the features of effective project cost management.
- Describe the features of effective project value management.

The past 30 years have witnessed a dramatic increase in the number and variety of organizations engaged in project-based work. In addition to "traditional" project-oriented industries, like construction, aerospace, and pharmaceuticals, service industries as diverse as finance, utilities, telecommunications, and insurance are beginning to embrace project-based ventures.

This paradigm shift is due to growing recognition that projects and their effective management can provide organizations with a significant competitive edge through cost reduction, enhanced responsiveness, and overall value to customers. Consequently, a number of organizations have adopted many of the well-known techniques of project management, and professional project management organizations have witnessed marked increases in membership.

Despite this enormous interest in projects and project management practices, success rates in many industries are at alarmingly low levels. In addition, bad news about high-profile projects continues to dominate the headlines—in both the public and private sectors. Consider these recent examples:

London's Costs for 2012 Soar. A British parliamentary committee criticized the spiraling costs of the 2012 London Olympics and called for greater transparency on finances. In November, Olympics Minister Tessa Jowell said infrastructure costs had risen by $1.8 billion from the $4.7 billion figure quoted in the bid. Some British lawmakers have speculated the total cost could reach more than $15.9 billion.[1]

Lockheed F-35 Joint Strike Fighter Cost Overruns. Lockheed Martin Corp. (LMT)'s F-35 Joint Strike Fighter, the Pentagon's most expensive weapons program, may face cost overruns of as much as 15 percent on early production models, U.S. Vice Admiral David Venlet said.

The plane's expense may exceed the contracted target cost by $964 million in the worst-case scenario, Venlet said in an April, 2011 interview in Washington. The overrun is calculated on $6.43 billion in aircraft and engine costs for 28 planes, according to F-35 program data. The low end of the overrun estimate is 11 percent, or $707 million, he said.[2]

Clearly, something is going wrong.

1.1 IMPORTANCE OF COST AND VALUE MANAGEMENT IN PROJECTS

The key features that define project success are twofold: managing costs to achieve efficiencies, and creating and enhancing value. These two elements enable project stakeholders to understand the activities and resources required to meet project goals, as well as the expenditures necessary to complete the project to the satisfaction of the customer.

Unfortunately, in the field of project management today, significant cost and schedule overruns are the norm, rather than the exception. In fact, research that examined the success rates of information technology (IT) projects indicates that the majority of these projects neither met their cost objectives nor delivered the promised value. For example:

- In a study of 300 large companies, consulting firm Peat Marwick found that 65 percent of hardware and/or software development projects were significantly behind schedule, were over budget, or failed to deliver value in terms of expected performance.[3]
- In a 2001 report on the current state of IT project implementation, the Standish Group predicted that out of a total of 300,000 projects that

Importance of Cost and Value Management in Projects

cost over $350 billion, approximately 43 percent will overshoot their initial cost estimates, while 63 percent will fall behind schedule and perform at only two-thirds of their expected capability.[4] In other words, these projects will meet neither their cost nor their value objectives.

Why do these problems persist, despite the fact that tools for cost efficiency and value enhancement are widely used, and their benefits are well understood? One key answer is the lack of an integrated cost and value management framework.

Before we explore this integration of cost and value, a brief discussion of their concepts in relationship to projects is worthwhile. Both require well-defined and structured management processes, commonly referred to as cost and value management. Project cost management focuses on issues such as cost estimation and budgeting, cash flow management, and cost control. On the other hand, the emphasis of value management is on optimizing project value—given cost, time, and resource constraints—while meeting performance requirements such as functionality and quality.

Cost and value management remains a critical but often underrepresented issue for a couple of reasons. First, in this book, we define value as the relationship between meeting or exceeding the expectations of project stakeholders, as well as the resources expended to meet or exceed those expectations. This definition clearly implies that project cost and value are inextricably linked, to the point where any attempt to enhance project value without a thorough understanding of its impact on cost and associated trade-offs is meaningless.

Second, project value is a multidimensional concept. Different project stakeholders with different vested interests have different perceptions about what constitutes value to them. For example, the expectations of top management often leave IT project teams scrambling to complete projects as quickly as possible. Internal customers, however, may request additional features that will delay completion. Each stakeholder sees value in the finished project; however, the measures they use to determine value can actually conflict. And yet, despite these differences, the one constant in any attempt to enhance project value is its cost ramifications.

The inability to clearly understand this complex relationship between project cost and value is one of the primary reasons why it is an underrepresented issue. The following case example illustrates this point.

4 COST AND VALUE MANAGEMENT IN PROJECTS

Case Study: Boston's Central Artery/Tunnel Project
The Central Artery highway in Boston was first opened in 1959 with considerable fanfare. Hailed as a technical marvel and model of proactive urban planning, the elevated six-lane highway was designed through the middle of the city and was intended to handle a traffic volume of 75,000 vehicles a day. However, by the early 1980s, the highway was overburdened by a daily volume of over 200,000 vehicles. Consequently, the city of Boston experienced some of the worst traffic congestion in the country, with bumper-to-bumper traffic that lasted for over 10 hours every day. The traffic woes of the Central Artery highway were further exasperated by an accident rate that was over four times the national average. Clearly, the Central Artery had not only become inadequate to handle the city of Boston's growing traffic volume, but had also become one of the most dangerous stretches of highway in the country.

To alleviate the problem, the City of Boston, under the supervision of the Massachusetts Turnpike Authority and with the help of Federal and State funding, came up with the Central Artery/Tunnel (CA/T) project, more commonly referred to in the Boston area as the "Big Dig." The two main features of the CA/T project are (1) an eight- to ten-lane underground expressway replacing the old elevated roadway, with a 14-lane, two-bridge crossing of the Charles River; and (2) extension of I-90 by building a tunnel that runs beneath South Boston and the harbor to Logan Airport. The CA/T project that began in the city in the early 1980s has been a work in progress for nearly 20 years.

From the outset, the CA/T project faced enormous technical and logistic challenges. First, the project involved construction of eight miles of highway with a total of 161 lane miles, with almost half them to be constructed underground. The project at its peak required 5,000 workers, excavation of 16 million cubic yards of soil, and 3.8 million cubic yards of concrete. Second, all of these construction activities had to be performed without disrupting existing traffic patterns, the current highway system, and its traffic flows.

The project began in 1983 with an original completion date of 1998 and a budget of $2.5 billion. However, neither the original budget nor the completion date has been met, and both have been revised upward frequently. For example, the original budget of $2.5 billion was adjusted to $6.44 billion in 1992, and $14.63 billion in 2003.

Because of the soaring cost projections and schedule overruns, the CA/T project has been source of considerable controversy. The situation was so bad that in 2000 a Federal audit of the project declared the Big Dig officially bankrupt. One of the audit's significant conclusions was that the out-of-control costs were due primarily to management's failure to hold contractors accountable for bids or mistakes. In fact, the public dissatisfaction over the delays and rising costs was so intense that the project manager of CA/T project had to resign. After more than 14 years of construction, the CA/T project was officially declared completed in the spring of 2006, in spite of the fact that some finishing work still remained. All of the tunnels and bridges and their connections and ramps to surface roads were opened to the public.

Unfortunately, the story does not end there. On July 10, 2006, the bolts holding four sections of cement ceiling panels (weighing 12 tons) failed, causing a section to collapse onto traffic below, where it tragically killed a commuter. Subsequent analysis of ceiling bolts in the rest of the tunnels showed 242 others already showing signs of stress, which led to a lengthy shutdown of the entire tunnel system for inspection and repairs.

Coupled with the March 2006 demand by the Massachusetts Attorney General for $108 million in refunds from contractors for "shoddy work" (including substandard concrete throughout the tunnel system), this event highlights the cloud under which the most expensive highway project in American history operates. Most recently, the State of Massachusetts formally assumed control of the Boston CA/T from the Turnpike Authority, and a concerted effort to pinpoint the causes of the Big Dig's poor cost estimation and control has begun.

In the final analysis, the Big Dig was certainly a technological marvel. It will undoubtedly provide enhanced value to its users through significant reduction in traffic congestion, carbon monoxide emissions, and improving the "green" reputation of the city. From a public relations and cost perspective, however, the project is currently viewed as a disaster.

Today, the taxpaying public continues to justifiably ask the fundamental question: "Where is the value in a project that has gone on for many years past its due date—and threatens to continue disrupting the lives of commuters in the Boston area?" The Massachusetts Turnpike Authority (MTA) believes that the answers lie in poor cost management and lack of adequate oversight from project managers.[5]

This book was written to explore the dynamic relationship between project value and cost, as well as the mechanisms used to achieve integration between them. Our goals are fourfold: first, to provide practicing managers with a thorough understanding of the various dimensions of cost and value in projects, the factors that impact them, and the most effective managerial approaches for achieving cost efficiency and value optimization. Unlike most project management books, which deal with this topic from a tactical perspective, this book takes a strategic approach.

Second, the book thoroughly covers the various elements of value management from a project perspective, including planning, engineering, and analysis. In addition, we examine various project management decision areas that have the potential to enhance value, along with relevant managerial approaches that can be used to optimize that value. Third, we provide an integrated framework for managing cost and value that can be useful to practicing project managers. Finally, this book contains a good deal of prescriptive advice on how to avoid common pitfalls in managing cost and value in projects.

We'll begin by exploring the essential features of effective cost and value management.

1.2 KEYS TO EFFECTIVE PROJECT COST MANAGEMENT

Effective project cost management is an extremely complex process that begins very early during a project life cycle, and long before its actual start. Among the factors that influence success is a reasonable and accurate system for estimating costs. Table 1.1, drawn from Rodney Turner's work,[6] highlights some of the most important considerations when creating a cost estimation system.

Table 1.1 Keys to effective cost estimation

1. A clear, complete, and unambiguous definition of the project and the scope of work involved
2. A thorough assessment of the potential risks involved, with well-thought-out action plans to minimize their possible impact
3. A well-trained and competent project manager
4. A thorough understanding, by all stakeholders, of the various types of costs that are likely to be incurred throughout the life of the project
5. A project organizational culture where there is a free flow of communication, so that all project participants clearly understand their responsibilities
6. A well-defined project work structure where work packages are broken down into manageable sizes
7. Meaningful budgets, where each work package is allocated its appropriate share of the total budget, commensurate to the work involved
8. An accounting system and coding scheme that are well aligned with the work breakdown structure and are compatible with the organization's management information system
9. A cost accounting system that will accumulate costs and allocate them to their relevant cost accounts as and when they are incurred
10. A prioritized and detailed work schedule, drawn from the work breakdown structure, which assigns and tracks the progress of individual tasks
11. Effective management of well-motivated staff, to ensure that progress meets or beats the work schedule

Table 1.1 Continued

12. A mechanism for comparing actual and planned expenditures for individual tasks, with the results extrapolated to cover the entire project
13. The ability to bring critical tasks that are late back on schedule, including providing for additional resources or taking other prompt remedial measures
14. Adequate and effective supervision to ensure that all activities are done right the first time
15. Supervision of staff time sheets so that only legitimate times are booked to various cost codes
16. Proper drafting of specifications and contracts
17. Discreet investigation to confirm that the customer is of sound financial standing, with sufficient funds to make all contracted payments
18. Similar investigation, though not necessarily as discreet, of all significant suppliers and subcontractors (especially those new to the contractor's experience)
19. Effective use of competitive tendering for all purchases and subcontractors to ensure the lowest cost commensurate with quality and to avoid committing to costs that exceed estimates and budgets
20. Proper consideration and control of modifications and contract variations, including charging all justifiable claims for price increases to the customer
21. Avoidance of all nonessential changes, especially those for which the project customer will not pay
22. Proper control of payments to suppliers and subcontractors to ensure that all invoices and claims for progress payments are neither overpaid nor paid too soon
23. Recovery of all incidental expenses allowed for in the contract charging structure (for example, expensive telephone calls, special printing and stationery, travel, and accommodation)
24. Proper invoicing to the customer, ensuring that claims for progress payments or cost reimbursements are made at the appropriate times and at the correct levels, so that disputes that could delay payments do not arise
25. Effective credit control to prevent payments from the customer from becoming long overdue
26. Internal security audits to help prevent losses through theft or fraud
27. Regular cost and progress reports to senior management, highlighting potential schedule or budget overruns in time for corrective action to be taken
28. Cost-effective design, perhaps using value engineering
29. Prompt action to close off accounts at the end of the project, to prevent unauthorized time bookings and other items from being charged to the project

While this list is not all-inclusive, its elements do have a significant influence over the effectiveness of cost management for projects large and small. Of more immediate interest is the sheer complexity and breadth of an effective cost estimation system, suggesting that organizations intent on controlling their costs need to recognize that there is no such thing as a simple, quick fix. Rather, downstream project cost management rests heavily on the care and accuracy of detailed estimation occurring early in the project. As the old saying suggests, "We can't fix what we can't see." Taken one step further, we can't control what we did not plan for!

1.3 ESSENTIAL FEATURES OF PROJECT VALUE MANAGEMENT

"Project value" refers to the relationship between the needs of different project stakeholders and the resources used to satisfy them. What constitutes project value, however, can be hard to pin down, because different stakeholders have different views. The challenge of value management is to understand and reconcile these differences.

Essentially, **value management** focuses on enhancing project value, given cost and time constraints, without any negative impact on the project facility's functionality, reliability, or quality. Effective project value management includes the following key features:

1. *Careful analysis and identification of project needs and scope*—The first step in effective value management lies in a clear project definition and scope analysis.
2. *Thorough planning of the project and subsequent work*—Effective planning helps to ensure that the project is developed to maximum cost efficiency with no unnecessary steps or wasted effort.
3. *Identification of key areas of opportunities that can influence project value*—The project team's goal is to enhance positive features of the project while keeping control of costs. Carefully weighing the cost/benefit analysis for a project is the key to enhancing value.
4. *Development of alternatives for exploiting the identified opportunities for improving project value*—Multiple paths may be available for enhancing value, some of which are more cost effective than others.

5. *Evaluation of alternatives; development of proposals and action plans*—Performing a clear trade-off analysis can help create alternatives and select the best choices for improving value.
6. *Use of a performance monitoring system for tracking project value*—The project team must have the means to accurately monitor the project, gain timely and actionable information, and make "on-line" decisions and choices among alternatives.
7. *Ensure a free flow of communication that cuts across organizational boundaries*—To effectively manage for value, a cross-functional mindset must prevail throughout the organization, so that ideas, alternatives, and creative solutions have the widest possible arena for exploitation.

In the remainder of this chapter, we will present an overview of the various topics that are discussed in this book.

1.4 ORGANIZATION OF THE BOOK

The book's key content and order of chapters are presented below, with Figure 1.1 providing the model for organization. We deliberately organized the book to resemble a project activity network, to illustrate the chronological sequencing of critical cost and value management activities, as well as the manner in which they must be integrated. This approach enables us to provide guidance regarding the important elements in these activities, and to offer some suggestions for a reasonable order in which to address them. Our ultimate aim is to propose an integrated framework by which project managers can incorporate cost and value thinking into their management style.

The opening chapter has explained the importance of this topic, reviewed the book's organization, and provided examples of the nature and scope of the problem, as well as its impact on the larger economy. Chapter 2 focuses on why it is critical to identify the needs and scope of a project at the earliest possible stage. This enables a project activities framework to be developed and project performance to be monitored—both of which ensure that promised value is delivered. An accurate definition of project scope and needs also sets the stage for identifying resources and approximate costs for successful project completion. Finally, needs assessment can alert project stakeholders, very early in

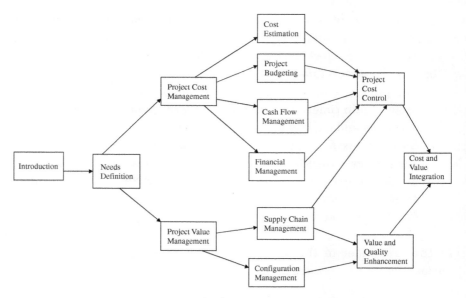

Figure 1.1   Cost and value management framework

the project life cycle, that the project as assigned may not meet needs and should be modified or canceled.

The next three chapters explore the topic of project cost management. Although the actual steps involved in the cost management process will depend on the nature of the project, the most commonly used sequence consists of cost estimation, budgeting, and cost control.

Chapter 3 covers **cost estimation**, which is important for a number of reasons. Estimation creates a standard against which actual expenditures can be compared, which in turn provides the basis for cost control. In addition, comparing cost estimates with estimates of return provides the basis for assessing project feasibility. When cost estimates are considered along with projected returns, they facilitate decision-making relating to project financing and funding. Cost estimates also provide mechanisms for managing cash flow during the course of a project, and for revising project activity duration—and they provide a framework for allocating resources as the project progresses. For example, as activity times are estimated on the basis of work content and resource availability, cost estimates for crashing project activities will help determine overall project acceleration costs. All of this is merely scratching the surface—we examine cost estimation in much greater detail throughout the course of the chapter.

Any discussion of project cost management must also include forecasting, which takes place at the front end of the project environment. **Forecasting** is the primary mechanism for ensuring that the project is on course, or for changing its direction if conditions in the project environment change. The context of our discussion relates to accurately forecasting activity times, as well as the relationship between project time/cost (also know as an S-curve), as they both have cost implications. Various approaches to forecasting—as well as some techniques that can be useful for forecasting activity times and the project S-curve—are found in the appendix to Chapter 3.

In Chapter 4, we discuss various concepts and approaches relating to **project budgeting**, which involves a combination of cost estimation, analysis, repetitive work, and, to a certain extent, intuition. An effective project budget is essentially a plan that integrates the allocation of resources, the project schedule, and the project's goals. Meaningful budgets require intense communication among all interested stakeholders, make use of multiple sources of data, and are developed concurrently with the project schedule to ensure that all milestones can be met.

When it comes to project cost control, a frequently made mistake is to compare the predicted rate of expenditures over time with the actual costs incurred. Without factoring in some measure of the work completed for expenditures incurred, any comparison for the purposes of cost control will lead to erroneous conclusions. The **earned value management (EVM)** method solves this problem. In essence, EVM requires that not merely the planned expenditure and actual cost incurred are measured, but also the *value of the work actually accomplished at the cost rates set out in the original budget*. In addition, the information EVM provides about the efficiency with which budgeted money is used relative to realized value makes it possible to create forecasts about the estimated cost and schedule to project completion. The topic of project cost control, along with the concepts of EVM and S-curves, are examined in detail in Chapter 5.

In the next two chapters, we discuss two other topics that can have a direct influence on overall project costs: cash flow management and financial management. **Cash flow management** is a vital process that is inextricably linked to the project cost management process—without it, effective cost management cannot be achieved. Simply defined, cash flow is the net difference between the flows of money in and out of the project. Cash flow management is important from the perspectives of both the project sponsor and its contractors. The sponsor must have access to funds to pay the contractor, and must be able to keep track of expenses and project progress. The contractor, on the other hand, should accurately monitor costs as they are incurred, bill the sponsor at the appropriate time, and have appropriate control systems to ensure that payments are received without excessive delays. The topic of managing cash flow is the focus of Chapter 6.

Project financing and its management is a vast subject area that cannot be examined in a single chapter. However, Chapter 7 covers some very important elements relating to the features and sources of project financing, and includes some key issues to consider in managing project finances. These are important elements of the overall project strategy—the ways in which a project is funded and its finances are managed have a significant impact on project cost, cash flow, and, most importantly, success. Sometimes, these outlays can be enormous and extremely risky: Boeing, for example, routinely commits well over $1 billion dollars in nonrecoverable R&D expenses for new project ventures. If the parent organization borrows the finances needed for the project, the money has to be repaid—regardless of success or failure.

The next series of chapters deal with how to manage and enhance value in projects. The topics include value management, change control and configuration management, project supply chain management, and quality management.

Chapter 8 explores the topic of **value management** in projects. Value management (VM) is a management style that focuses on enhancing overall business performance through innovation, skills development, and people motivation, and by creating synergies. Value management encompasses all value-enhancing techniques, including *value planning, value engineering,* and *value analysis.* In this chapter, we discuss issues such as the relationship between value management and value engineering, value planning, and value analysis, the difference between value and cost management, and various VM techniques.

The essence of **configuration management (CM)** is to gain the continual agreement of stakeholders regarding configuration and functionality of the project's end product throughout the project's lifecycle. By doing so, value to the project's stakeholders is enhanced. Specifically, configuration management is the process by which the configuration of individual components manages the functionality of those components, as well as the functionality of the total product. Configuration management is also used to manage the work methods by which each component is made and built in to the product. We will learn about CM and its various features in Chapter 9.

Chapter 10 covers **supply chain management (SCM)**, a process that attempts to fully integrate the network of all organizations and their related activities in an efficient manner. The focus of SCM is to add value to the product or service at each stage of the chain, so that it meets or exceeds customer expectations. (For this reason, supply chains are often referred to as "value" chains.) While it has enjoyed great popularity elsewhere, SCM is only recently gaining ground among many project organizations. Consequently, project organizations not employing supply chain processes are continually plagued by problems of poor quality, low profit margins, and schedule and cost overruns. The reality is that project organizations with their multitude of suppliers have an urgent need to adopt SCM-related practices. To that end, we discuss the unique nature of project supply chains in this chapter, as well as some strategies for applying SCM-related concepts and techniques to projects.

The concept of "quality" is multidimensional. What constitutes quality, and how quality is defined, may vary among different customer groups. For example, among piano buyers, performing artists who have the knowledge and sophistication to evaluate the subtle differences between the piano's tone and feel are more likely to buy a piano such as the Steinway Grand. Less sophisticated consumers, who are generally more interested in consistent quality, will more likely buy a Yamaha or Baldwin upright piano.[7] In our chapter on value management, the concept of project quality is dealt with indirectly under the umbrella term of "value." Similarly, the quality dimensions of performance and functionality are addressed in the material on configuration management. However, given the uniqueness of a project-based environment, the role of quality and its contribution in enhancing the overall value of projects takes on added importance. We examine the various aspects of a project's quality and quality management in Chapter 11.

The final chapter is organized in a capstone format that provides an integrated framework incorporating the entire scope of this book. Our discussion focuses on the required processes that should be in place to effectively manage cost, time, value, and risk aspects of a project. For example, we explore how risk management approaches can be used to protect project value from inevitable uncertainties.

As we have said, one of the greatest challenges facing project managers today is the need to develop a better understanding of the roles that cost and value management play in successful project implementation. Project management texts touch the surface of these duties, but it is a rare work that gives them their due. This is a pity, particularly given the numerous projects that fail due to the inability of project teams and organizations to effectively manage these twin requirements.

Readers will find that we offer a sequential methodology for addressing cost and value management. First, it is necessary to effectively plan for them. Just as we plan for a project's development through sophisticated means such as work breakdown structures and network diagrams, we must include effective cost and value planning in our project management repertoire. Second, cost and value must be controlled on an ongoing basis. Projects, as we know, do not succeed or fail solely on the basis of sound planning—project execution requires us to apply those plans in a meaningful way. Likewise, cost and value management include a definite "control" process to monitor the ongoing status of cost and value and make necessary corrections. Finally, good project management also requires an evaluation stage, in which we analyze what went right and what went wrong. For cost and value management, the analysis component is equally critical. It is only through critical analysis that we learn. It is only through learning that we improve.

We believe that this book comes at an opportune time in the advancement of project management knowledge and processes. In recent years, projects have become one of the key means by which organizations add to their bottom lines—making it critical to master project-based skills, including cost and value management. Proficiency in these two areas will confer an important advantage on organizations competing with each other for profits, market share, and technical innovations. We hope that readers find much of "value" for themselves and their organizations in these pages.

REFERENCES

1. London's costs for 2012 soar. MailOnline, August 2, 2011.
2. Lockheed gets Navy warning shot. *Wall Street Journal*, January 16, 2007, p. A4.

3. Pinto, J. K., and Millet, I. (1999) *Successful Information Systems Implementation: The Human Side*, 2nd ed. Newtown Square, PA: PMI.
4. Standish Group (2001) *Extreme Chaos*. Boston, MA: The Standish Group International.
5. Pinto, J. K. (2009) *Project Management: Achieving Competitive Advantage*, 2nd ed. Upper Saddle River, NJ: Pearson Prentice Hall, p. 248.
6. Turner, J. R., and Turner, R. (2008) *Gower Handbook of Project Management*, 4th ed. Aldershot, Hampshire, UK: Gower, pp. 295–297.
7. Garvin, D. A. (1992) *Operations Strategy: Text and Cases*. Upper Saddle River, NJ: Prentice Hall, pp. 127–128.
8. Kapur, G. K. Technology Leadership Institute (1997) Survey conducted at Brookings Institute, December 2–3, Washington, DC.

KEY TERMS

Project value
Value management
Cost estimation
Forecasting
Project budgeting
Earned value management (EVM)
Cash flow management
Project financing
Configuration management (CM)
Supply chain management (SCM)

Chapter 2

Project Needs Assessment, Concept Development, and Planning

LEARNING OBJECTIVES

- Define corporate and project strategy and understand the difference between the two.
- Describe the critical steps in developing an accurate project concept.
- Describe Statement of Work (SOW) and understand its key features.
- Describe the Work Breakdown Structure (WBS) and its development.

The value that a project delivers within an organization must be evaluated in terms of its contribution to strategic goals—if the project fails to provide value, it won't survive. This link between corporate and project strategy is so critical that project value must be understood in this context.

For our discussion, corporate strategy is defined as the framework for articulating an organization's overall goals and objectives, along with the means that will be used to accomplish them. Project strategy, in contrast, outlines various internal and external factors that will affect project success, as well as the processes and approaches that will be adopted to attain project objectives.

From an organizational point of view, projects invariably evolve out of corporate strategies. The corporate strategy is typically cascaded through several strategic business units, and is ultimately translated into a collection of program or project portfolios. Given available resources, these projects are the mechanisms through which corporate initiatives are implemented.[1]

In a project environment, all project planning is part of the larger umbrella process called project strategy. Therefore, linking project strategy to the corporate strategy ensures that project plans are not in conflict with corporate goals and objectives. Project planning, however, cannot be initiated unless there is a clear understanding of what the project is expected to accomplish, along with its overall scope. For this reason, project needs identification and scope definitions are fundamental prerequisites to managing a project's cost and value.

> **Case Study: Evaluating Project Opportunities**
> Project-based organizations typically evaluate all project opportunities on a variety of criteria, usually including some notion of strategic "fit," or alignment with corporate goals. For example, among the criteria Weyerhaueuser Corporation uses to evaluate new project opportunities are (1) significant changes in the external environment; (2) long-term future needs of lead customers; (3) business strategies, priorities, and technical needs; and (4) corporate strategic direction.
>
> Hoechst Pharmaceuticals, The Royal Bank of Canada, Exxon/Mobil, Pfizer, and 3M are just a few of the organizations that recognize that effective projects must be directly aligned with corporate strategic objectives, and routinely screen all opportunities to verify this strategic fit. In all cases, regardless of the industry or "type" of project under consideration, a project will not be undertaken unless it is expected to contribute to and support corporate goals—no matter what the profit potential might be.[2]

Projects typically follow a life cycle, as shown in Figure 2.1. This life cycle consists of four stages: needs identification and conceptual development, project planning, project execution, and project termination. This model will be an important concept as we explore the main theme of cost and value management. This chapter focuses on the first two stages in the project life cycle as they relate to the critical activities in identifying project needs and developing appropriate and detailed plans to address those needs. Later chapters will continue to explore the manner in which we manage projects for value as they progress through the life cycle toward completion.

Needs Identification

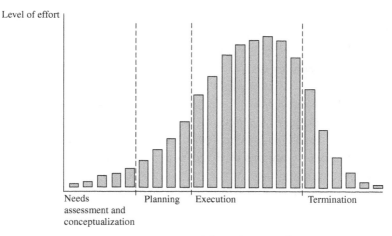

Figure 2.1 Project life cycle stages[3]

2.1 NEEDS IDENTIFICATION

The impetus to launch a project can originate either from the project organization or from an external customer. In both cases, **needs identification** is the first stage of the project life cycle. Projects initiated by the project organization can stem from the need to solve an underlying problem, or to exploit an opportunity in the external environment. In a similar fashion, an external customer who has identified a problem, or is dissatisfied with status quo, sees some benefit in launching a project that can result in substantial improvements over the existing condition. Needs identification from the external customer typically leads to the search for a contractor who can carry out the project, as well as the issuance of a **request for proposal (RFP)**.

Regardless of the project's source of origin, the one important principle in the needs identification stage is *an unambiguous, clear definition of the problem or need so that the best possible solutions to the problem or the best possible approaches to meet the need can be found.* The process of clear problem definition or needs identification may require extensive data collection to determine the problem's nature and magnitude. Once this has been accomplished, it is possible to establish project goals and objectives.

During the project needs assessment phase, the needs of *all* project stakeholders must be clearly defined, with no ambiguity. At this juncture, it is not important to determine whether the project can satisfy all these needs, or to worry about the best approaches to meet them. All we are attempting to do during this stage is to identify the various and often-conflicting expectations of the different stakeholders.

When the needs of the project are clearly understood and defined, several benefits can accrue:

- First, it becomes possible to determine all project activities necessary to accomplish the desired results.
- Second, a sound basis is provided for monitoring and tracking project performance to ensure that the real needs identified for the project are, in fact, being met.
- Third, the project organization can be alerted that the project as conceptualized may not meet the needs of the interested parties, and may have to be modified or canceled.
- Fourth, the project organization is enabled to assess the importance of each need in terms of value delivered to the project as a whole, as well as the relative cost of each need. Armed with this vital information, it is possible to conduct a formal cost-benefit or cost-value analysis to determine whether or not to proceed with the project.

With all of this information in hand very early in the project life cycle, organizations are able to create a solid foundation for managing both the cost and value of the project. For example, let's assume that a project team has identified ten needs for an information technology (IT) project. The task at hand is to prioritize these needs in terms of their contribution to cost and value of the project. Table 2.1 lists these ten needs and their corresponding impact on cost and value, expressed as percentages. As you can see, 100 points are distributed among the different needs, based on assumptions as to how well they are likely to affect project cost and value.

Next, the team generates a reasonable estimate of total project costs that will be derived from satisfying the ten identified needs. It is determined that the effort will cost a total of $150,000, with the most expensive need, number five, estimated to cost $43,500. As a percentage of the overall costs, satisfying this need will have a 29 percent cost impact on the project. Likewise, successfully addressing project need number four, labeled "25 percent customer satisfaction improvement within three months," is estimated to be the second most costly issue to resolve (20 percent of total costs, or approximately $30,000). However, successfully addressing this need will also have a significant positive effect through improving the project's value.

Needs Identification

Table 2.1 Cost impact and value contribution of project needs

Project need	Cost impact in percentages	Value impact in percentages
1. Cost reduction	7	3
2. Upgrade legacy software to new software system	6	7
3. Eliminate coding rework	7	4
4. 25% customer satisfaction improvement within 3 months	20	12
5. 10% lead-time response reduction within 6 months	29	33
6. 10% IT downtime decrease within 3 months	4	20
7. Streamline help desk response	6	10
8. Reduce billing errors by 20% within 3 months	8	5
9. 25% cycle time improvement within 6 months	9	6
10. Vendor selection	4	0

Figure 2.2 is a plot of the data presented in Table 2.1. It can be clearly seen that project need number six (10 percent IT downtime decrease within three months) falls into the high-priority category, while needs one and ten are of low priority. The rest of the needs fall into the medium-priority category. It is interesting to note that project need number five (10 percent lead-time response reduction within six months) not only has the highest cost, but also delivers the highest value. In light of this, the project team needs to make a decision about the importance of this need to the overall project, as well as whether it's worth addressing, despite its high cost.

Beyond understanding and identifying project needs, a project team must establish that the needs are, in fact, significant or have merit, and can be met at a reasonable, practical cost. This formulation will allow project managers and their teams to make reasonable cost/value trade-off decisions that are likely to have the greatest overall impact on the project. Once this analysis has been completed, we can consider the next stage of the project life cycle: conceptual development.

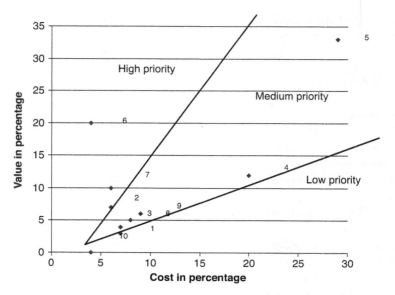

Figure 2.2   Cost-value analysis of project needs

2.2   CONCEPTUAL DEVELOPMENT

The purpose of **conceptual development** is to pare down the complexity of the project to a more basic level. This process begins by focusing on project objectives and searching for the best possible ways to meet those objectives.[4]

To develop an accurate project concept, the project management team collects key pieces of data and information. The following are the critical steps in this process[5]:

- *Statement of goals*—In this step, the project team establishes that a need or problem exists that requires a solution, as well as how the project intends to address the need or solve the problem.
- *Gather information*—When gathering relevant pieces of data for the project, the search must be thorough and complete, and the project team must explore all available sources of information. This is a vital step in initiating the project, as it provides a clear picture of the existing state of affairs, possible supply sources, sources of funding, extent of top management support, and specific target dates for the project.
- *Constraints*—The project team should anticipate and must have a thorough understanding of the potential limitations that could adversely impact project development. These include constraints in terms of time, budget, or customer demand.

- *Alternative solutions*—The problem or need that the project is attempting to address typically has many alternative solution approaches. Searching for these alternatives prevents the project organization from initiating a project without first ensuring that more efficient or effective options exist. The search for alternative approaches requires that the nature of the problem is clearly understood. This, in turn, enables the project team to get a clearer understanding of the project's characteristics, including its unique features, and provides a variety of ways for initiating and implementing the project. During this process, it is very likely that the project team will be presented with an innovative or novel project development alternative.
- *Project objectives*—Conceptual development concludes with a clear statement of the final objectives for the project in terms of value that will be delivered, approximate cost that will be incurred, required resources, and expected completion time.

All of the steps in the conceptual development process function together as components of a system that ultimately affects the project's value and cost. If each of the steps is properly executed, then the achievement of project objectives, in particular, the cost and value objective, will logically follow.

To completely and accurately address a problem or need, the project team must explicitly define it. The problem statement should include a concise description of project goals and objectives that can be easily understood by the project team. If a project is initiated without this, it will inevitably experience cost overruns and schedule delays.

Let's assume, for example, that an organization is attempting to address a very simple problem of streamlining the link between the billing function and end-of-month reporting. If the goal of this IT project is vaguely defined as "improving billing and record-keeping operations," it is possible that the IT department could misinterpret the goal and launch a project that provides a complex solution with multiple interactive screens, costly user retraining, and the generation of voluminous reports. In this case, poor problem definition could lead to a relatively simple solution being replaced by one that is unnecessarily complex and expensive.

A well-defined problem statement will provide the project team with a thorough understanding of the problem to be addressed—which, in turn, will pave the way for project value optimization, cost reduction, and eventual project success. In addition, it will provide a convenient reference point that the team can revisit should problems occur during project development.

2.3 THE STATEMENT OF WORK

The **statement of work (SOW)** is a detailed narrative of the work required to complete a project.[6] The document clearly delineates the objectives and requirements of the project, and identifies key activities in broad terms. A useful SOW also contains expected project outcomes and any funding or schedule constraints.

The statement of work often triggers the initiation of a project, with the performance and design requirements typically providing the basis for subsequent contractual agreements. Once the SOW becomes contractual, it is used to evaluate contractor performance. It also serves as a standard for evaluating the proposals of competing contractors in terms of whether they can meet stated performance requirements.

A statement of work moves from the general to the specific, beginning with background that includes a brief history of the reasons why the project is needed. Then, it identifies component tasks before moving to a more detailed discussion of each task objective and the approach necessary to accomplish it. A comprehensive SOW includes the following considerations:

- It defines *all* of the work to be completed, but without a high level of detail. This will enable the potential contractor to estimate probable project cost, levels of expertise needed, and resources required to complete the requirements.
- It includes design and performance requirements in both qualitative and quantitative terms, but does not describe specific technical requirements. For example, in an IT project, a SOW may instruct a contractor to develop, implement, test, and maintain a system, but will not state specifically how it should be done.
- It establishes, either directly or indirectly (by referencing other documents), the nonspecification requirements expected.
- It sets forth performance requirements for the contractor's management in terms of expected results, although the exact procedures to achieve them are omitted.
- It establishes the criteria that will be used to evaluate the project as major phases are completed. Once this is accomplished, the document delineates all key tasks and subtasks of the project to be completed, including a narrative description of the work required.
- It clearly states the specific duties of the contractor, with no uncertainty as to whether or not the contractor is obligated to perform

certain tasks. In addition, a SOW should include only the minimal specifications and standards relevant to the task, and should be tailored to limit costs. Also, only the documents needed to satisfy existing requirements should be referenced. There should be a clear demarcation between the background information and suggested procedures and the responsibilities required of the contractor.
- It should include a brief scope statement of what is included and excluded from the SOW. Under no circumstances does the scope section include directions to perform the work, specification of data requirements, or description of the deliverable products.
- It typically includes a section on documents. This section contains a list of only those documents referenced in the work requirements. While this section is initially left blank, it is updated whenever documents are identified for inclusion. Unless otherwise stated, total compliance with the documents listed is absolutely required—improper document referencing can be a major source of cost escalation.[7]

Table 2.2 is a standard template for the construction of a reasonably detailed SOW for most projects. Obviously, the unique characteristics of individual projects may require that you provide greater or lesser degrees of detail.[8]

Table 2.2 Elements in a comprehensive statement of work

Date Submitted	
Revision Number	
Project Name	
Project Identification Number	
SOW Prepared by:	

1. DESCRIPTION AND SCOPE
    a. Summary of Work Requested
    b. Background
    c. Description of Major Elements (Deliverables) of the Completed Project
    d. Expected Benefits
    e. Items Not Covered in Scope
    f. Priorities Assigned to Each Element in the Project
2. APPROACH
    a. Major Milestones/Key Events Anticipated

Table 2.2 Continued

Date	Milestone/event

    b. Special Standards or Methodologies to be Observed
    c. Impact on Existing Systems or Projects
    d. Assumptions Critical to the Project
    e. Plans for Status Report Updates
    f. Procedures for Changes of Scope or Work Effort
3. RESOURCE REQUIREMENTS
    a. Detailed Plan/Rationale for Resource Needs and Assignments

Person	Role and rationale

    b. Other Material Resource Needs (Hardware, Software, Materials, Money, etc.)
    c. Expected Commitments from Other Departments in Support
    d. Concerns or Alternatives Related to Staffing Plan
4. RISKS AND CONCERNS
    a. Environmental Risks
    b. Client Expectation Risks
    c. Competitive Risks
    d. Risks in Project Development (Technical)
    e. Project Constraints
    f. Overall Risk Assessment
    g. Risk Mitigation or Abatement Strategies
5. ACCEPTANCE CRITERIA
    a. Detailed Acceptance Process and Criteria
    b. Testing/Qualification Approach
    c. Termination of Project
6. ESTIMATED TIME AND COSTS
    a. Estimated Time to Complete Project Work
    b. Estimated Costs to Complete Project Work
    c. Anticipated Ongoing Costs
7. OUTSTANDING ISSUES

2.4 PROJECT PLANNING

The next stage of the project life cycle is **project planning**. At this stage, we assume that a contractor or project organization has been chosen to develop and implement the project.

Two of the most important aspects of project planning are **project scope definition**, which sets the stage for developing a comprehensive project plan, and the development of the **work breakdown structure (WBS)**. Scope definition establishes the criteria that will be used to evaluate whether the major phases of the project have been completed. The outcome of this process is to develop a document called the **work scope document**, which describes project objectives and requirements.

The terms "project scope" and "statement of work" are often used synonymously, and there are many similarities between the two. However, there is an important difference. While the SOW document is developed by the project client, the scope document is developed by the contractor or project organization responsible for developing and implementing the project. Once the client chooses the contractor for the project, the SOW document can be used as the basis for further refinement, typically as a result of negotiations between the contractor and the client. The outcome of this process is the project scope document, which serves as the foundation for all downstream project development activities.

A full and complete definition of the project scope leads to the next most important step in project planning: the development of the **work breakdown structure (WBS)**. In this phase, the overall project is broken into manageable and well-defined work packages. The creation of a WBS ensures that no element of work is overlooked or duplicated, and that the relationships among work packages are clearly identified and understood.

The WBS is the basis on which project schedules and cost estimates are developed, and also provides the framework for assigning management and task responsibilities. While we will discuss WBS in more detail later in this chapter, it is important to note that a well-defined project scope and an accurate WBS are the twin pillars for ensuring project success. The two, in tandem, constitute the basis for effective project cost and value management.

2.5 PROJECT SCOPE DEFINITION

The first major endeavor of the project planning team is to identify, in broad terms, the principal activities to be undertaken. While elaborate detail is not required, the range of activities must encompass all of the work that needs to be completed, as well as all of the major participants. Subsequently, a document is developed that identifies and describes all the principal tasks and subtasks to be completed, with a complete description of the work required. This is the **project scope document**.

2.5.1 Purpose of the Scope Definition Document

The purpose of the scope definition document is to provide the project organization and the project manager with a road map of both the work to be completed, as well as the types of final deliverables sought. In some aspects, the scope document and the SOW are practically identical. For example, while the scope document may describe the end product or service to be produced and delivered by the project, it should not be treated as a document for technical specifications. (Technical specifications documents are developed independently and separately from the scope document, and are subject to frequent revisions, particularly for innovative or developmental projects.)

In addition, while information such as the target delivery date and target budget are included, the main purpose of the scope document is to thoroughly and comprehensively define the work to be completed and the expected deliverables. Typically, the scope document presents schedule requirements at some broad level that may only identify the starting and ending dates, as well as any major milestones.

2.5.2 Elements of the Scope Definition Document

Several key elements must be included in a scope document. Like the SOW, the scope document also includes project objectives, deliverables, milestones, technical requirements, and limits and exclusions. In addition, the scope document should also include a section for reviews with customers. We will now examine these elements in detail, along with several aspects and issues involved in developing a scope document.

The scope document should clearly and explicitly describe the objectives of the project, generate a detailed, comprehensive list of all deliverables, and delineate all work requirements. This list enables the project team to determine the nature of the work to be performed, the project team members responsible for specific work, and the expected outcome at completion.

Because the scope document is developed during the early stages of the project, it will undergo extensive revisions and will be frequently updated throughout the life cycle of the project. It is very likely that the scope of the project may be diminished or enhanced, and, as a consequence, the work content of the project may change. It is important to analyze and assess the impact of changes to the scope, so that the necessary adjustments to the project's schedule and budget can be made. It is also important to obtain necessary approvals *before* the project scope is changed, with all project stakeholders informed of the changes.

The scope document should also define the limits of the project's scope by clearly identifying any aspects that will not be the responsibility of the project organization or contractor. For example, in a house construction project, the installation of a concrete driveway may be included in the project scope, while landscaping may be excluded and left to the owner. Failure to identify those aspects can lead to false expectations or an unnecessary waste of time and resources.

The level of detail and description in the scope document depends on the customer, the project organization, and, to a great extent, the nature of the project itself. Returning to our example of the construction of a house, the scope document can be relatively simple and straightforward. On the other hand, for a project such as the $200 billion Joint Strike Fighter (JSF) program—one of the largest and most complex project management undertakings in history—the scope document is bound to be highly descriptive and complex.

A well-defined project scope document has often been cited as one of the most critical project success factors. Consequently, great care should be taken in its development. Some of the common shortcomings seen in poorly designed scope documents include the following:

- Lack of clear distinction among tasks, standards, specifications, and approval procedures
- Lack of a coherent structure, pattern, or chronological order
- Failure to fully describe the tasks to be completed

- Use of confusing, vague, overly general, or incorrect language
- Failure to receive customer review

When these occur, the undesirable consequence is total misinterpretation of the project scope, by both the customer and other subcontractors. This inevitably leads to disputes and misunderstandings over what needs to be done and what is excluded in project development.

Table 2.3 is an example of a scope document for the construction of a house. As you go through it, you can see examples of the scope document elements discussed in this chapter.

2.5.3 Project Scope Changes

As we've said, changes to the original project scope occur frequently over the course of a project. They may happen for any number of legitimate reasons, including a technical breakthrough that renders the original goals obsolete; a desire to alter the project's functionality or cut features to save money; the discovery of unanticipated physical difficulties, such as poor soil properties or groundwater at a construction site; and so forth. Customers who want additional features often initiate scope changes, as do contractors.

Regardless of their causes, scope changes remain a seemingly inevitable phenomenon that must be addressed and managed if project cost overruns, schedule delays, and even failure are to be avoided. These negatives can be greatly minimized if the original scope document contains clear, precise performance requirements, task descriptions, and resource requirements that can be understood and agreed to by the various project stakeholders. For these reasons, many project organizations have developed detailed manuals for preparing scope documents.

Because the scope document is developed during the early stages of project planning, it is impossible to foresee all of the work or the extent of work that may be needed later on in the project. Therefore, during scope document preparation, it is important to completely list all of the assumptions that were made, and include it as an appendix to the document. In the event that an assumption is rendered invalid during the course of the project's progress, necessary amendments to the scope document should be made.

Table 2.3 Scope document for a house construction project[10]

PROJECT OBJECTIVE

To construct a high-quality, custom home within seven months at a cost not to exceed $250,000.

DELIVERABLES

- A 2400-square-foot, 2-bath, 4-bedroom, finished home
- A finished garage, insulated and sheet rocked
- Kitchen appliances to include range, oven, microwave, refrigerator
- High-efficiency gas furnace with programmable thermostat
- Air conditioning is included

MILESTONES

1. Permits approved—March 5
2. Foundation poured—March 14
3. Dry in. Framing, sheathing, plumbing, electrical, and mechanical inspections passed—May 25
4. Final inspection—June 7

TECHNICAL REQUIREMENTS

1. Home must meet local building codes.
2. All windows and doors must pass NFRC class 40 energy ratings.
3. Exterior wall insulation must meet an "R" factor of 21.
4. Ceiling insulation must meet and "R" factor of 38.
5. Floor insulation must meet and "R" factor of 25.
6. Garage will accommodate two large-size cars and a storage area.
7. Structure must pass seismic stability codes.

LIMITS AND EXCLUSIONS

1. The home will be built to the specifications and design of the original blueprints provided by the customer.
2. Owner responsible for landscaping.
3. Dishwasher is not included among kitchen appliances.
4. Contractor reserves the right to contract out services.
5. Contractor responsible for subcontracted work.
6. Site work limited to Monday through Friday, 8:00 a.m. to 6:00 p.m.

CUSTOMER REVIEW

Doug and Donna Smith

Within reasonable limits, a project manger can accommodate changes to the project scope as long as the project deadline and budget can be met. On the other hand, if there is any risk of exceeding the project completion date or budget, a more formal approach to managing scope changes is warranted.[9] In the event that such a scope change is deemed imperative, the customer and all project stakeholders should be consulted, and a formal review should take place before the change is made. To effectively manage this process, it is absolutely essential to define the responsibilities of the various project stakeholders during the preparation of the scope document, and to identify the person responsible for making the necessary decisions at the appropriate time.

2.6 WORK BREAKDOWN STRUCTURE

A project team armed with a well-defined project scope document can proceed to the next most important phase in project planning: development of the **work breakdown structure (WBS)**. The WBS is a hierarchical structure in which all of the major phases of project work are organized in a logical manner. Simply stated, it is a layout of how the work of the project ought to be performed and managed.

The WBS links together the project scope, the schedule, and the cost elements, and forms the basis for the entire project information structure. It provides the framework that captures the project's work information flows, as well as the mechanism for tracking the project's schedule and progress. It also provides the basis for summarizing and reporting the project's cost data.

In essence, the WBS creates a common framework that

- Presents the entire project as a series of logically interrelated elements
- Facilitates project planning
- Provides the basis for estimating and determining project costs and budgets
- Provides the mechanism for tracking project performance, cost, and schedule
- Provides the basis for determining the resources needed to accomplish project objectives
- Provides the mechanism for generating project schedule and status reports

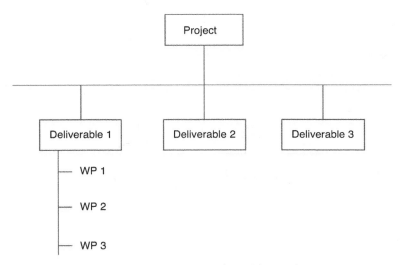

Figure 2.3 Diagram of WBS hierarchy

- Triggers the development of the project network and control systems
- Facilitates the assignment of responsibilities for each work element of the project

Graphically, we can imagine a WBS along the lines shown in Figure 2.3. There are some important features to note here. First, the breakdown of project activities is hierarchical, moving from the overall project level, down through specific project deliverables (and sometimes subdeliverables), and finishing with work packages. According to the Project Management Institute's PMBoK, work packages are the elemental tasks at the lowest level in the WBS and represent basic project activities for which duration and budget figures can be assigned.

Along with a well-defined project scope document, a WBS with sufficient levels of detail is an invaluable tool for managing scope change. Project scope can easily be expanded or diminished by including or eliminating the appropriate work packages, and by making the necessary adjustments in terms of resources and schedules.

2.6.1 Types of Work Breakdown Structures

There are many different types of work breakdown structures. The **product-based WBS** is most suitable for projects that have a tangible

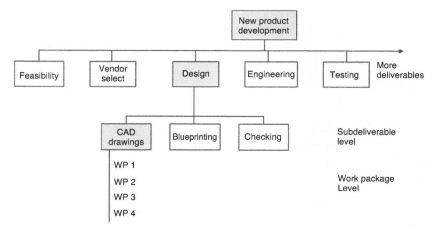

Figure 2.4 Product-based WBS

outcome, such as the design and construction of a tunnel, or a new fighter jet prototype. In these kinds of product-based work breakdown structures, it is relatively easy to break the project into deliverables, subdeliverables, and ultimately into manageable work packages. A product-based WBS also facilitates cost estimation of the various product components, which in turn provides a basis for comparison with the project's actual cost performance. An example of a product-based WBS is shown in Figure 2.4.

In an **organizational breakdown structure**, known as OBS, the organizational units responsible for completing the work requirements of the project are integrated into the structure. The clear advantage of an OBS is that it establishes, in a hierarchical format, accountability, responsibility, and reporting relationships for the various organizational units that are involved in completing the work. An example of an OBS is shown in Figure 2.5.

When a project involves a series of phases or tasks, such as in an information systems project, a **process-oriented WBS** is more appropriate. Unlike a product-based WBS, a process-oriented WBS involves completion of major tasks in sequence as the project evolves over time. In general, a process-oriented WBS can be better integrated with the overall project, because the tasks can be delineated to the lowest level, while the product-based WBS is most useful for accumulating product cost information. In addition, hybrid structures that involve a combination of features from both product- and process-based structures can be used. An example of a process-oriented WBS is shown in Figure 2.6.

Work Breakdown Structure

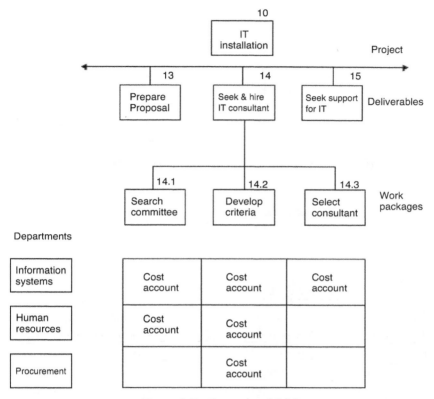

Figure 2.5 Example of OBS

In the present-day project environment, with its proliferation of computerized project management systems, both types (product and process) of WBS can be used. However, the choice really depends on the particular project, its purpose, the organization's information system and how it interfaces with the other organizational units, and, ultimately, how useful it is as a planning aid.

2.6.2 Work Breakdown Structure Development

The construction of the WBS is worthy of special consideration, because it is critical to the information structure that will be set up to cover most of the project's managerial data set. In this section, we review the procedure for constructing a WBS, as well as some of the practical aspects that should be considered in developing one.

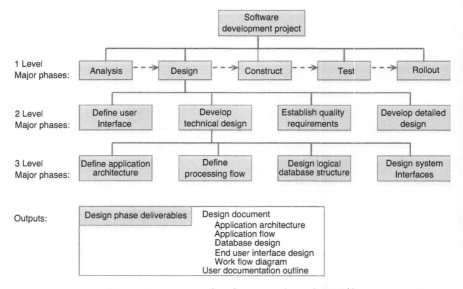

Figure 2.6   Example of process-based WBS[11]

Constructing a WBS is a group effort. The responsibility for the top three levels usually lies with the project manager, who typically works with other managers to develop the major deliverables found there. At the lower levels, however, input into the details should come from those responsible for the day-to-day work. In addition to the project team, each person responsible for a specific work package must be consulted.

There are no universal standards as to the number of levels to use in a WBS. This is typically dictated by the complexity of the project, and the number and nature of the activities. Most have three or four levels at a minimum, although more may be needed if there are significantly more subdeliverables and work packages.

As a rule, the greater the number of work packages, the greater the amount of time, effort, and cost incurred in developing the WBS. However, the increased number of packages also significantly enhances the accuracy of project status monitoring and reporting. On the other hand, larger but fewer work packages reduce the cost of WBS development, but the accuracy of project status monitoring and reporting suffers. In most practical situations, a compromise on the number and size of work packages is struck.

The most important levels in the WBS development process are the top few levels, which represent the project and associated major deliverables, and the lowest levels, which represent the work activities that are to be scheduled and resourced. Typically, the intermediate levels are an aggregation of the lower-level activities, and serve as the mechanism for reporting. The intermediate levels can also be used to set up cost accounts through which cost and budgetary information can be gathered and monitored.

A WBS should be constructed by separating the total work, in a hierarchical framework, into discrete and logical subelements that reflect completeness, compatibility, and linkage to the project's end item. This hierarchical structure facilitates the aggregation of budget and actual costs incurred into larger cost accounts. In this way, the total value of the work at one level is an aggregation of all the work completed one level below, which enables project work and cost performance to be monitored. The WBS should also reflect all of the project requisites, as well as functional requirements, and should incorporate both recurring and nonrecurring costs.

In a WBS, the lowest level consists of tasks of short duration that have a discrete start and end point. Because they consume resources, costs are often reported at this work package level. Work packages serve as control point mechanisms for monitoring and reporting performance, in terms of whether associated tasks were completed according to specifications, on time, and within budget. The responsibility for completing work packages should be assigned to the appropriate organizational units and individuals, and the resulting WBS should reflect the reporting requirements of the organization.

We can develop a WBS by one of two methods: top-down or bottom-up.

- The **top-down approach** begins with the total work required to produce the project's end product. This is successively broken down, first into major work blocks and then into more detailed work package levels until a required level of detail has been achieved.
- The **bottom-up approach** begins with a comprehensive list of all the activities or tasks required to complete the project. Then it moves up, in a hierarchical format, first to work packages, then to cost accounts, and so on, until the top level represents the total work to be performed.

It is highly unlikely that the first version of the WBS will be the last. More often than not, it will undergo frequent revisions in response to feedback that work must be broken into more detailed tasks. Clearly, only individuals who have a thorough knowledge of the work to be performed can conduct the appropriate level of detailed breakdown. In addition, it is impossible to establish estimates of task duration, resource requirements, and the network of precedence relationships among activities without complete agreement on details at the lowest level. One final caveat: while the WBS is very much a part of the overall project planning process, the two data sets for project planning and the work breakdown structure are distinct entities and should be kept separate.

2.6.3   Coding of Work Breakdown Structures

Correctly coding a WBS facilitates the identification and definition of its various levels and elements. Coding schemes also enable consolidation of reports at any level of the WBS, which makes them useful for tracking budgetary and cost information. While a variety of schemes can be used, the most common method is numeric indention.

For example, in the product-based WBS shown in Table 2.4, code 1040 indicates the step of seeking and hiring an IT consultant, while one of the work packages for this deliverable, 1041, requires that the company form a search committee. This kind of coding process is repeated for each WBS element.

Just as the hierarchical levels become more specific as we move down, project coding typically follows a similar degree of specificity. For example, in the task-based WBS for a hypothetical MIS installation project shown in Table 2.4, code 1000 designates the final deliverable, while code 11200 represents the lowest manageable subdeliverable. Other features that can be incorporated in a WBS include the use of alpha characters and additional numeric characters to indicate the type of work, geographical location, vendor type, and so on.

2.6.4   Integrating the WBS and the Organization

The final phase of WBS development is integrating it with the various organizational units. This is done to identify the organizational units responsible for performing the work, to provide a framework for summarizing performance, and to provide a link between the organizational unit and cost control accounts. The end result of this process is the aforementioned organizational breakdown structure (OBS),

Table 2.4 Example of a WBS for a MIS installation project

Breakdown	Description	WBS Code
MIS Installation Project		**1000**
Subproject 1	**Match MIS to organizational tasks and problems**	**1010**
Activity 1	Conduct problem analysis	1011
Activity 2	Develop information on MIS technology	1012
Subproject 2	**Identify MIS user needs**	**1020**
Activity 1	Interview potential users	1021
Activity 2	Develop presentation of MIS benefits	1022
Activity 3	Gain user "buy-in" to system	1023
Subproject 3	**Prepare informal proposal**	**1030**
Activity 1	Develop cost/benefit information	1031
Subproject 4	**Seek and hire MIS consultant**	**1040**
Activity 1	Delegate members as search committee	1041
Activity 2	Develop selection criteria	1042
Activity 3	Interview and select consultant	1043
Subproject 5	**Seek staff and departmental support for MIS**	**1050**
Subproject 6	**Identify the appropriate location for MIS**	**1060**
Activity 1	Consult with physical plant engineers	1061
Activity 2	Identify possible alternative sites	1062
Activity 3	Secure site approval	1063
Subproject 7	**Prepare a formal proposal for MIS introduction**	**1070**
Subproject 8	**Develop RFPs from vendors**	**1080**
Activity 1	Develop criteria for decision	1081
Activity 2	Contact appropriate vendors	1082
Activity 3	Select winner(s) and inform losers	1083
Subproject 9	**Conduct a pilot project (or series of projects)**	**1090**
Subproject 10	**Enter a contract for purchase**	**10000**
Subproject 11	**Adopt and use MIS technology**	**11000**
Activity 1	Initiate employee training sessions	11100
Activity 2	Develop monitoring system for technical problems	11200

40 PROJECT NEEDS ASSESSMENT, CONCEPT DEVELOPMENT, PLANNING

which shows how the firm has organized to discharge work responsibility. The intersection of work packages and organizational units generates cost accounts that will be used later for project control purposes.

An example of how WBS levels and tasks can be linked with budget and personnel is shown in Table 2.5.

Table 2.5 Cost and personnel assignments

WBS Code	*Budget*	*Responsibility*
1000	700,000	Stevie Rocco, MIS Manager
1010	5,000	Peg Thoms
1011	2,500	Peg Thoms
1012	2,500	Dave Winfield
1020	2,750	David Spade
1021	1,000	David Spade
1022	1,000	Demi Moore
1023	750	Ken Griffey
1030	2,000	James Garner
1031	2,000	James Garner
1040	2,500	Susan Hoffman
1041	-0-	Susan Hoffmann
1042	1,500	Susan Hoffman
1043	1,000	Kim Basinger
1050	-0-	Ralph Nader
1060	1,500	George Bush
1061	-0-	John Kerry
1062	750	Sammy Sosa
1063	750	Barry Bonds
1070	2,000	Bob Williams
1080	250	Beth Deppe
1081	-0-	Kent Salfi
1082	250	James Montgomery
1083	-0-	Bob Williams
1090	30,000	Debbie Reynolds
10000	600,000	Jeff Pinto
11000	54,000	David Spade
1110	30,000	David Spade
1120	24,000	Michael Jordan

2.6.5 Guidelines for Developing a Work Breakdown Structure

At this point, it should be clear that there are no universal standards or rules that can be applied to developing a WBS—which can make it a challenging and daunting task. However, the following guidelines may be useful in many situations[12]:

1. Each element of a WBS, at any level, should be tied to the deliverables at that level. In this way, there will be no ambiguity about whether an activity has been completed (product delivered) or whether it is still ongoing (product not yet delivered).
2. Each element of a WBS at a higher level should be broken down into more than one element at the immediate lower level. If this does not happen, then one of the levels is redundant. The only exception is when upper levels are mandated by procedural requirements.
3. Each element of a WBS at the lowest level must be assignable as an integral unit to a single department, vendor, or individual. This is the first part of the answer to the question, "What level of detail should exist within the WBS?" If no one department, vendor, or individual is responsible for an activity, then not only may there be accounting and scheduling problems, but no one is likely to assume responsibility for it.
4. The riskier elements of the WBS should be broken down into greater details to enable the risks to be identified and dealt with. This is the second part of the answer to the question, "What level of detail should exist within the WBS?"
5. In developing the WBS, it is important to distinguish between it and the project plan. As both the work breakdown structure and the cost structure feed information into the overall project plan, these three entities are interrelated—and any changes to any one of them may have an impact on the others.

In closing, the project management concepts discussed in this chapter are of paramount importance to effective cost and value management. Without a precise definition of project needs, a well-defined SOW and project scope documents, and an accurate and well thought-out WBS, neither value optimization nor efficient management of costs in projects is possible.

REFERENCES

1. Jamieson, A., and Morris, P.W.G. (2007) Moving from corporate strategy to project strategy. In P.W.G. Morris and J. K. Pinto (Eds.), *The Wiley Guide for Managing Projects.* Hoboken, NJ: Wiley, pp. 177–205.
2. Pinto, J. K. (2009) *Project Management: Achieving Competitive Advantage,* 2nd ed. Upper Saddle River, NJ: Pearson Prentice Hall.
3. Pinto, J. K., and Rouhaianen, P. (2002) *Building Customer-based Project Organizations.* New York: Wiley.
4. Laufer, A. 1991. Project planning: timing issues and path of progress. *Project Management Journal,* 22 (2); Stuckenbruck, L. C. (1989). *The Implementation of Project Management: The Professional Handbook.* Boston, MA: Addison-Wesley.
5. Pinto, J. K. (2009) *Ibid.*
6. Martin, M. G. (1998) Statement of work: the foundation for delivering successful service projects. *PM Network,* 12 (10).
7. (n.d.) http://home.btconnect.com/managingstandard/sow_1.htm
8. Pinto, J. K. (2009) *Ibid.*
9. Mocha, T. (2002) "Project managers can accommodate scope changes—to a point." http://techrepublic.com.com/5100-10878_11-1027711.html
10. Gray, C. F., and Larson, E. W. (2011) *Project Management: The Managerial Process,* 5th ed. New York: McGraw-Hill Irwin.
11. Gray, C. F., and Larson, E. W. *Ibid.,* Figure 4.8, p. 117.
12. Adapted from UMIST Module 3 workbook, p. 3.04.23.

KEY TERMS

Needs identification
Request for proposal (RFP)
Conceptual development
Statement of work (SOW)
Project planning
Project scope definition
Work breakdown structure (WBS)
Work scope document
Project scope document
Product-based WBS
Organizational breakdown structure
Process-oriented WBS
Top-down approach
Bottom-up approach

Chapter 3

Cost Estimation

LEARNING OBJECTIVES

- Explain the importance of accurate cost estimation.
- Discuss the problems that cause cost overruns.
- Identify sources and categories of project costs.
- Demonstrate methods for estimating project costs.

Horror stories of project failures due to unprecedented cost overruns are reported in the business press on almost a daily basis. Of these, London's Millennium Dome serves as a classic case study.

> **Case Study: London's Millennium Dome**
> One of the central features of London's millennium celebration was the creation of the Millennium Dome, a project that invoked strong "love it/hate it" feelings from day one. Aesthetic controversies aside, when it comes to poor cost estimation and misguided optimism, the Millennium Dome stands supreme.
>
> Built on the site of a former gasworks in Greenwich, southeast London, the 20-acre, circular, tent-like building is twice the size of the Georgia Dome in Atlanta and has the largest roof of any structure in the world. In 1996, when the British government announced plans for the Dome, it was intended to be a celebration of British achievement and progress, with each of its 14 interior zones represented by titles such as "Learning" and "Journey." Although similar in theme to Walt Disney World's Epcot Center in Florida, the Dome offered audiences the opportunity to meander at their own pace, rather than offering structured rides.
>
> Upon their ascension to power in 1997, the Labor government of Tony Blair seized on the project with enthusiasm, suggesting that its enormous

cost could be safely funded through national lottery money. By the time the Dome was completed, the total cost was estimated at $1.2 billion dollars, including the construction of a public transportation line (parking at the site could accommodate only one percent of daily projected visitors), and other ancillary support services. The British government was not unduly worried, as it estimated that the 12 million visitors needed to break even would quickly swarm to the site, making the Millennium Dome one of the most visited and talked-about attractions in London.

Government projections for cost and revenue never worked out as they had hoped. In the month the Dome opened, a mere 364,000 people visited (about three percent of the breakeven figure required). Responses from the public were strongly negative, as they complained about long lines and attractions that did not justify the high prices and lack of amenities.

By the end of June 2000 (halfway through the first year's operations), the Dome had attracted three million visitors, quite a difference from the projected (and needed) 12 million. Even after downward revisions to a new projection of seven million visitors, the Millennium Dome continued to lose money at an astounding rate. Losses, by midsummer, had already approached $150 million dollars, requiring a cash infusion from the British government to keep the doors open.

What's to happen at the site of the now-vacant Millennium Dome? Officially renamed the "O_2" by a consortium that recently purchased it, the Dome has been proposed as a venue for indoor sporting events, including a plan to host the 2009 International Gymnastics Championships. This proposal, however, is not without considerable political opposition; already, there are calls to convert the 300-acre site into thousands of houses to alleviate some of London's congestion. Perhaps out of urban development, some positive results will emerge. The same cannot be said for the Millennium Dome, a mixture of bad cost control and appallingly optimistic projections.[1]

The Millennium Dome's lesson is that *all* projects, regardless of size, require meticulous planning and a thorough analysis of expected project costs, along with the associated value delivered. To that end, this chapter focuses on the issues and methods associated with estimating the costs of projects.

3.1 IMPORTANCE OF COST ESTIMATION

A recent survey conducted on information technology (IT) projects revealed that more than 83 percent fail to meet their economic goals. Two of the major culprits cited for IT project problems are **erroneous cost estimates** and **overlooking key project costs**.[2]

Estimating project costs are important for a number of reasons:

- They provide a standard against which actual expenditures incurred during the course of a project can be compared, and serve as the basis for cost control.
- They are the chief means for assessing project feasibility. A comparison of the cost estimates with the estimates of revenues will enable the project organization to determine if the project is worthwhile to undertake.
- Along with project returns, they facilitate decisions relating to project financing and funding.
- They provide the mechanism for managing cash flow during the course of the project.
- They give the project manager a framework for allocating scarce resources as the project progresses.
- They provide the mechanism for revising project activity duration.

To a great extent, project success hinges on accurate cost estimation. While a number of factors can contribute to cost overruns, one of the main culprits is project acceleration. As we will show in Chapter 4, this is because the relationship between cost and time is not always linear. When a project crashes, relevant costs combine in a nonlinear fashion and cause a vicious cycle where they increase several times more than what was originally anticipated. This phenomenon underscores the importance of developing methods for accurate cost estimation that take into account uncertainties, the "portfolio effect," and the various dynamics of project crashing.[3]

Cost estimation and project budgeting are inextricably linked. For example, identifying the various human and material resources needed for a project and developing accurate cost estimates for them are essential parts of developing a comprehensive, time-phased project budget, as well as for subsequent project monitoring and cost control. Without reasonable cost estimation, project budgets are essentially useless, and without accurate budgeting, cost estimation is a wasted exercise.

3.2 PROBLEMS OF COST ESTIMATION

Most project managers would agree that cost overruns are the causes of more frustration and anguish than almost any other factor. Clearly, this is the case with innovative development projects, as the very nature of innovation makes it nearly impossible to accurately predict costs. However, even in relatively routine projects (such as many found in the construction industry), initial cost estimates are often completely off target from the final outcome.

Unfortunately, industry data suggest that overruns are the norm, rather than the exception. The most significant reasons for this are

- Low initial cost estimates
- Unanticipated technical difficulties
- Lack of or poor scope definition
- Specification changes
- External factors

Low initial cost estimates are often the result of underestimating the magnitude and complexity of the task to be undertaken. The obvious reason for this is that the evaluation of project task performance and duration is often done in isolation, without considering the impact of surrounding activities. In addition, we incorrectly assume that everything will go as planned, and fail to anticipate problems that may unexpectedly surface in the future.

Other factors that lead to low initial cost estimates are corporate attitudes to internal expenditures, as well as business or political gamesmanship. For example, project managers may feel that presenting an initial low cost estimate will increase their chances of gaining board level approval for their project, making it more likely to win in a competitive situation. Another common, erroneous view is that if a low initial budget is set, a lower overall project cost will result than if a higher and more reasonable figure is fixed.

The unfortunate effect of these factors is that they lead to the approval of projects that have no sound basis. One of the best examples of a project in which political expediency led to unrealistic initial cost estimates is the Channel Tunnel (or "Chunnel") that connects England and France. While the Chunnel itself was an outstanding project from a technical point of view, it was a financial disaster plagued with significant cost overruns.

The problem of low initial estimates can be overcome as uncertainty is reduced, and as project managers become more knowledgeable and gain confidence in the cost estimation process. Because estimation is ultimately part of the entire cost management cycle, the more familiar a project manager becomes with the cycle as a whole, the more accurate his or her initial cost estimates will be.

Unanticipated technical difficulties are the second important cause of cost overruns. The roots of this problem usually rest in poor initial design, but this is not always the case. Quite often, the testing stage will generate insights that may not have been readily apparent in the initial design stage, typically in the form of problems and technical failures that must be resolved. When these issues surface, there is no other recourse than incurring additional costs, because meeting technical specifications is fundamental to product performance and acceptability.

An example of a project that encountered unforeseen technical difficulties is the wobbling Millennium Bridge across the Thames River in London. The bridge, which was built at a cost of over $25 million, started swaying uncontrollably and was closed within days of its opening in June 2000. The cause of the problem was a building phenomenon known as "synchronous foot fall" that the designers were not aware of, even though it was investigated and its effects were reported in 1993. It was estimated that additional expenditures of up to $15 million dollars could be incurred to solve the synchronous foot fall problem and reopen the bridge to traffic.

In this example, the failure to recognize that technical problems are inevitable with innovative design projects resulted in an expensive lesson. Meaningful cost estimates, in contrast, should factor in the potential for technical problems, start-up delays, or other technical risks.

Lack of or poor scope definition leads to the creation of projects that have no clear direction, features, goals, or even purpose. It is important to recognize that when the initial steps of developing a comprehensive scope statement and work breakdown structure are done poorly, they effectively turn any attempt to reasonably estimate project costs into an exercise in futility.[4] Examples of this problem exist in many of the pure research and development programs undertaken by private and public organizations. Many times, scientists and research engineers become fixated on the study of a task for its own sake, losing sight of the larger picture or goals under which the research was taken.

Specification changes, often referred to as **"scope creep,"** make initial cost estimates nearly meaningless, and are often the primary culprits for overruns. Requests for specification changes while the project is underway sometimes originate internally within the project organization, but more often than not, the sources for these requests are external.

A number of factors can necessitate specification changes during the project development process, including dynamic changes in the market, newly perceived requirements from customers, and new legislative mandates related to the product. For example, requests for additional features, significant modifications, and updated processes are quite common (and frequent) in information technology projects, particularly after project activities based on the original scope are initiated.[5] Regardless of the reasons, the undesirable consequence of specification changes is that initial cost estimates have to be revised, because some of the project work already completed must be negated.

External factors such as inflation, interest rates, environmental issues, and currency exchange rate fluctuations can also escalate actual project costs, particularly in the case of projects where technical problems and other difficulties lead to an increase in project duration. In the case of multinational public sector projects, delays due to politics among the nations involved can result in significant increases in actual project costs incurred.

The case of the Dulhasti Power project in India is a classic example of project cost estimates run amok. Originally conceived to tap into the vast potential of hydroelectric power, the project was a massive undertaking. It was sited in the remote foothills of the Himalayas, leading to a construction schedule that was disrupted by recurring geological surprises.

The project team was further burdened with terrorism concerns, political problems, and power struggles between state and central government over control and regulation of the facility. Finally, because foreign firms were originally selected to manage development, there were international cooperation problems to be sorted out as well.

As initially conceived, the project's cost was estimated at 1.6 billion rupees (about $50 million). By the time the contract was let, cost estimates had risen to 4.5 billion rupees and later, successively, to 8, 11, 16, etc. Today, the final price tag appears to be 170 billion rupees (nearly $5.3 billion) for a project that is 18 years late.

In addition to the above factors, there are other, "softer" issues associated with understating or overstating resource requirements. First, many managers fail to realize that estimates, by definition, are predictions, and are fraught with uncertainty. Consequently, the actual cost incurred for an activity may vary from the original cost estimate by a significant percentage (plus or minus). Second, inaccuracies in cost estimation typically occur at the time of project initiation, and their impact is most significant if the profitability of the project is in doubt.

Estimates become progressively more accurate as we progress through the project life cycle. By the time of major development, cost estimates invariably are more precise, as shown in Figure 3.1. However, the most

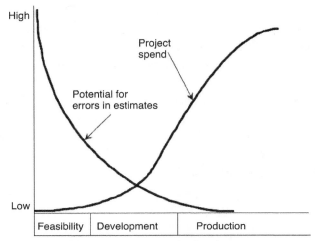

Figure 3.1 Potential for errors in estimates during the various stages of the project progress[6]

significant of all decisions—whether or not to start the project—is made at the very beginning. If the premise upon which that decision is based is false, the negative repercussions, both from cost and schedule perspectives, can be very serious.

3.3 SOURCES AND CATEGORIES OF PROJECT COSTS

Initial cost estimation begins during project proposal development. At this stage, all relevant costs that are likely to be incurred in the project should be identified and included in the initial project proposal. To develop the initial cost estimates, we need to know the various sources of project costs. These include

1. *Cost of labor*—This involves hiring costs and wages for the various human resources associated with the project. Given that a project requires a variety of personnel with varying skill levels, labor cost estimation is not an easy task. It must factor in salaries or hourly rates, pension and health benefits, other overhead, and an estimate of time involvement.
2. *Cost of materials*—This is the cost of raw materials, supplies, and other equipment needed to complete project tasks. The actual costs incurred for materials depend on the nature of the project; for example, material costs for software development projects can be quite small, whereas in construction projects they are very large.

3. *Cost of equipment and facilities*—Many "off-site" projects, such as mining or construction of large buildings, require project team members to rent facilities and equipment. In these cases, the rental costs are legitimate costs that can be charged against the project.

Other relevant sources of project costs can include travel costs of team members, as well as costs associated with subcontractors or consultants.

Project costs can be classified as direct or indirect, recurring or nonrecurring, fixed or variable, and normal or expedited. **Direct costs** can be directly charged against the project; for example, the costs of personnel who are directly involved in the project, or the costs of materials directly used for project work. **Indirect costs** include overhead, as well as selling and administrative expenses. Examples of overhead costs include costs associated with taxes, insurance, utilities, and so forth. Costs associated with selling and administrative expenses stem from salaries, commissions, advertising, etc. Tracking and allocating indirect costs to specific projects is considerably more difficult than allocating direct costs. Consequently, among project organizations there is a wide variation in the approaches employed to estimate and allocate indirect costs. Some organizations use a percentage multiplier (typically between 120 and 150 percent) for overhead costs, on top of direct costs.

Recurring costs, such as labor and materials, are repeatedly incurred throughout the project life cycle. **Nonrecurring costs**, on the other hand, are one-time costs that are typically incurred at the beginning or at the end of the project, such as market research and labor training. **Fixed costs** do not vary with usage. For example, costs incurred in the purchase of capital equipment remain fixed, regardless of the extent of equipment use. **Variable costs**, on the other hand, vary directly with usage. They are typically associated with labor and materials. **Normal costs** are incurred when project tasks are completed according to the original planned duration. **Expedited costs** or **crash costs** are unplanned costs incurred as a result of steps taken to accelerate project completion. For example, costs associated with using additional overtime or hiring additional workers specifically to hasten project completion can be regarded as expedited costs.

While all of the above are viable approaches to classifying project costs, it should be emphasized that many of these costs belong to multiple classifications; for example, labor costs can be regarded as direct, recurring, variable, and normal costs.

3.4 COST ESTIMATING METHODS

At times, estimating project costs seems to resemble an art form as much as a science. There are two important project "laws" at work regarding cost estimation:

- First, *the better we define the project's various costs in the beginning, the less chance there is of making serious estimating errors.*
- Second, *the more accurate our initial cost estimates, the greater the likelihood of preparing a project budget that accurately reflects the true budget for the project* and the greater our chances of completing the project within budget estimates.

One key is to cost out the project on a disaggregated basis; that is, to first break the project down by deliverables and work packages as a method for estimating task-level costs. These estimates can then be aggregated into an overall project budget. For example, rather than attempt to create a cost estimate for completing a deliverable comprising four work packages, it is typically more accurate to first identify the costs for completing each work package individually and then create a deliverable cost estimate (see Table 3.1).

Organizations use a number of methods to estimate project costs, ranging from highly technical and quantitative to more qualitative approaches. Among the more common cost estimation methods are the following[7]:

1. *Ballpark estimates*—These types of estimates, also known as "order of magnitude" estimates, are often used when there is not sufficient information or time available to develop more accurate or detailed estimates.

Table 3.1 Disaggregating project activities to create reasonable cost estimates

Project activities	*Estimated cost*
Deliverable 1040—Site preparation	
Work Package (WP) 1041—Surveying	$ 3,000
WP 1042—Utility line installation	15,000
WP 1043—Site clearing	8,000
WP 1044—Debris removal	3,500
Total cost for Deliverable 1040	$29,500

Typically, ballpark estimates are used for making competitive bids for project contracts, or for initial rough-cut estimates of resources needed for a project. For example, a company preparing an RFQ (request for quote) may develop a ballpark estimate of resource requirements to determine whether it is worthwhile to bid on the project with a more detailed analysis. As a general rule, ballpark estimates should attempt an accuracy of ±30 percent. Clearly, with such a large margin of error, ballpark estimates cannot replace more thorough, accurate, and detailed cost estimation.[8]

2. *Feasibility estimates*—These estimates are developed after preliminary project design work is completed. With the initial scope of work in hand, it is now possible to begin the process of developing the initial project baseline by sending RFQs to suppliers and other subcontractors. Feasibility estimates are often used in construction projects, where published information on material costs is widely available. This information, along with estimates of material quantities involved, facilitates fairly accurate cost estimates for a wide range of project activities. As more relevant information becomes available further down the project life cycle, the ±10 percent margin of error in feasibility estimates is more stringent than ballpark estimates.

3. *Definitive estimates*—These estimates can be developed only after the completion of most design work. At this stage, there is very clear understanding of the scope and capabilities of the project, and changes to project specifications are virtually nonexistent. Furthermore, a comprehensive project plan is in place, all activities and their sequence required for project completion have been identified, and all major purchase orders have been submitted based on known prices and availabilities of materials and equipment. Given that definitive estimates are developed further down the project life cycle with more accurate information and fewer project uncertainties, these estimates provide a much more accurate expected cost of the project at completion, with a ±5 percent margin of error.

4. *Comparative estimates*—Comparative estimates use historical data from previous project activities as the frame of reference for current estimates. One comparative estimating method is **parametric estimation**, which holds that most projects are similar to previous projects by way of similar features or parameters and, therefore, are likely to incur similar costs. For example, Boeing Corporation

routinely develops detailed estimates of current projects by taking older work and inserting a multiplier to account for the impact of inflation, labor and materials increases, and other reasonable direct costs. This parametric estimate, when carefully performed, allows Boeing to create highly accurate estimates when costing out the work and preparing detailed budgets for new aircraft development projects. Parametric estimation consists of two steps:

a. Identifying the features/parameters of an older or well-known project that can be directly related to the development cost of the current project; and
b. Defining the mathematical nature of that relationship.

Figure 3.2 provides an example of parametric estimation in the context of the development of the Concorde aircraft in the 1960s. The Concorde represented such a unique and innovative airframe design that it was difficult, if not impossible, to estimate the amount of time required to complete the airplane schematics. However, using parametric estimation based on experiences with other recently developed aircraft, a linear relationship was discovered between the number of man-weeks needed to design the aircraft and its projected cruising speed. Using these values, it was possible to make a reasonably accurate projection of the expected design budget. Parametric estimation demonstrated that, in spite of significant changes in airplane design over past decades, the relationship between cruising speed, size, and design effort held remarkably steady. Given the availability of high-speed computers, parametric estimation has the potential to improve the speed and accuracy of both the initial estimating and design evaluation processes. The method can be used in variety of project scenarios, including

- Estimating costs before a full-scale design
- Quickly choosing from alternate design schemes at the time of project initiation, when there is very little time to do a detailed design of each concept or determine cost by conventional methods
- Generating bids when time is limited to develop a full cost proposal
- Setting design-to-cost targets
- Evaluating the cost of changes to existing products
- Assessing the effects of new technology on product development and pricing

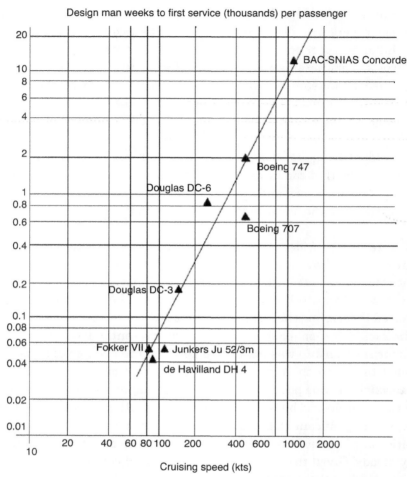

Figure 3.2 Parametric relationships between cruising speed, size, and design effort that has held good despite the remarkable changes in airliner technology from the DH 4 to Concorde[9]

Parametric Cost Estimation Example

AMEC Inc., a major firm in the industrial construction industry, was interested in developing a quick cost estimate for the construction of a warehousing facility. In the recent past, the company completed construction of ten similar warehousing projects with similar general architecture, layout, and construction materials. AMEC engineers analyzed the relationship between material costs and several building parameters, and developed a multiple regression model using the method of least squares (this method is discussed in the appendix to this chapter).

Based on existing historical data, AMEC understood that overall material costs for a new warehouse were directly related, in a linear manner, to the total floor space and number of loading docks planned for the facility. As a result, the multiple regression model relating material costs (Y) to floor space (X_1 in terms of 10,000 square feet) and number of shipping docks (X_2) in a facility is given below. That is,

$$Y = 315.41 + 47{,}319 * X_1 + 19{,}638 * X_2$$

All regression measures of model fit that were calculated indicated that the model would be a very good predictor for the data from the previous ten projects. Using the above model, engineers were able to estimate the material costs for a new warehousing facility with 500,000 square feet and five docks as follows:

$$Y = 315.41 + 47{,}319 * 50 + 19{,}638 * 5 = \$2{,}464{,}455.41$$

Given the variety of cost estimation approaches available, it is fair to ask which method project organizations should employ. The answer depends on knowledge of the firm's industry (e.g., software development versus construction), its ability to account for and manage most project cost variables, its history of successful project management, the number of similar projects completed in the past, the knowledge and resourcefulness of project managers, and the company's budgeting requirements.

In some instances (for example, extremely innovative research and development projects), it may be impossible to create cost estimates with better than a ±20 percent degree of accuracy. On the other hand, in events project management (for example, managing a conference and banquet), it is reasonable to prepare definitive budgets quite early in the project. The key lies in a realistic appraisal of the type of project being undertaken, the speed with which various cost estimates must be created, and the comfort level top management has with cost estimation errors. It would be foolhardy to provide cost estimates disguised as "accurate" when they were based on simple order of magnitude guesswork. Conversely, if the information is available, it is reasonable to expect the project team to provide as accurate a cost estimate as possible, as early as possible.

3.5 COST ESTIMATION PROCESS

Both initial and detailed cost estimates should be prepared with diligence. Even though insufficient information is available when initial estimates are prepared, it is imperative that these estimates are developed carefully, for several reasons. First, they serve as the basis for future release of funds to the project. Second, they become standards against which future estimates are compared. Third, they are used to determine likely overall project costs and return on investment. Above all, the decision as to whether or not to proceed with the project is often made on the basis of initial estimates. Because it is virtually impossible to foresee all project tasks at this point, it is a good idea to include some contingency, which is discussed in more detail later in this chapter.

3.5.1 Creating the Detailed Estimate

The proliferation of computerized project management systems in the past decade has placed a greater emphasis on detailed planning, and has changed the nature of cost estimation. By focusing on the details of each individual activity in the project plan, these computerized project planning systems can develop a project cost estimate, including the resources required and their associated charge rates for each activity. These detailed estimates are extremely useful for project control, because the individual tasks are often aggregated to work packages, and costed work packages are now a significant aspect of the entire project control cycle.

Computerized systems offer many advantages:

- As each project activity is examined in detail—including its ramifications on resources required and other associated tasks—the resulting cost estimate tends to be considerably more accurate. In addition, established ratios and historical data can also be used to further improve accuracy.
- Many project planning software packages can generate a time-phased cost schedule from the cost structure, project plan, and work breakdown structure (WBS). This cost schedule can be used as a task budget against which actual cost performance can be measured.
- From the perspective of project control, the most important advantage of computerized systems is their ability to generate a cost plan that is linked to the project work plan.

Cost Estimation Process **57**

However, despite these advantages, computerized methods have two important shortcomings. First, a meaningful cost plan cannot be developed unless a project plan with considerable detail is available. Second, as the estimates of resources and associated skills required for activities are often based on subjective judgments, the potential for inaccuracy is still high.

The cost estimation process presented in Figure 3.3 can be done entirely manually, which is the practice in many organizations. While many

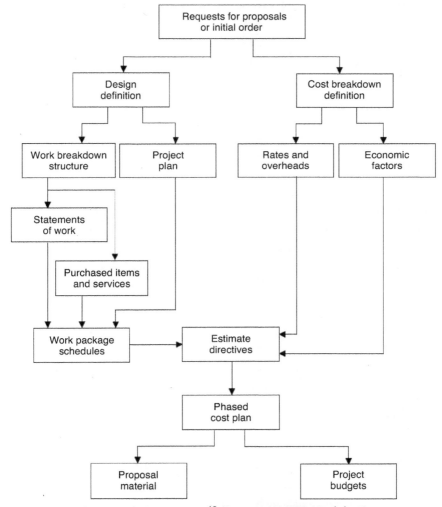

Figure 3.3 Cost estimation process[10] (Source: UMIST Module 4)

ESTIMATE AND QUOTATION SHEET				
Project No.	Description:			Type No.
Work Package No.	Task No.			Estimate No.
Work Package Description		Task Description		
Internal Labor				
Skill	Category	Rate	Hours	Cost
Senior Test Engineer	TE4	18.50	40	$740.00
Test Engineer	TE3	14.00	80	1120.00
Fitter	PF4	13.30	30	399.00
Drafter	DR2	15.00	15	225.00
Drawing checker	DR3	16.50	3	49.50
Sub-Total, Hours and Costs			168	2533.50
Labor Contingency (10) %			17	254.00
Total Labor, Hours and Costs			185	2787.50
Overhead (80) %				2230.00
Gross Labor Cost				5017.50
Bought-Out Costs				
Materials (Specify): Bolts plus cleating material				20.00
Finished Goods (Specify): N/A				
Services and Facilities: Hire test house; Instrumentation plus Report				12300.00
Sub-contract Manufacture (Specify): Fixture & Bolt Modification				250.00
Sub-Total				12570.00
Contingency (15) %				1885.50
Total Bought-Out Costs				14455.50
Expenses				
Specify: On-site accommodation plus traveling				340.00
Total Costs				14795.50
Profit (N/A) %				
Total Quoted Sum				14795.50
Compiled by:				
Approved			Date:	

Figure 3.4 Sample project activity cost estimating sheet

of the calculations can be computerized—including the computations of overheads, inflation rates, and contingencies—the initial estimate of resources and quantities required still requires human intervention. In situations where computer systems are not available, or where basic cost information has to be compiled, a standardized cost estimation sheet (such as the one is presented in Figure 3.4) can be used.

The above cost estimation sheet comprises four main sections. The top section provides information about the project, work package, and task for which the estimation is being done. The following three sections (in order) contain details of labor content, bought-out expenditures, and other expenses. It is important to keep the bottom three categories of cost separate, as different companies use different accounting conventions.

In addition to basic costs, labor and bought-out expenses have provision for a contingency element, while the expenses section lists items such as traveling, accommodations, special insurance, freight charges, and entertainment. While all of the costs in the sample presented above are compiled at current rates, cost revisions due to economic factors are not normally taken into account at the initial estimation stage.

3.6 ALLOWANCES FOR CONTINGENCIES IN COST ESTIMATION

The contingency element in the above cost estimation worksheet requires further explanation. Because estimates are predictions of future costs, there is always the potential for error, and cost overruns are more the norm than the exception. Consequently, additional allowances are needed to act as a buffer, so that the funds actually allocated to the project will be sufficient.

This allowance is known as **contingency**. It is an important aspect of project finance, because it provides a certain amount of protection against unknown and uncertain elements that can derail a project. As general rule, the greater the degree of uncertainty, the greater the amount of contingency required. The availability of contingency funds also enhances the viability and value of the cost estimation process as a control tool.

Sometimes, significant problems can occur that make it necessary to overspend even the contingency amount. In spite of this drawback, there are several benefits to including the contingency element in the cost estimation process. First, the approach recognizes that the future is uncertain, and that unexpected problems can increase overall project costs.

Second, contingency estimates make explicit provisions for potential and unexpected cost escalations in project plans and budgets. Finally, the use of contingency funds provides an early warning signal of potential cost overruns.

Example of Detailed Cost Estimation

Bradley Hunter is the project manager for Reliance Corporation's XYZ project. He begins preparing a project cost estimate by breaking the project into eight work packages, along with a preliminary schedule. For each work package, Bradley estimates the number of labor hours per week for each of the three labor grades assigned to the project. Table 3.2 presents the summary information on labor hours per week for each labor grade, labor costs, overhead rates, and other nonlabor costs. The costs for labor grades 1, 2, and 3 are $25, $30, and $40, respectively, and the overhead rates applied to these grades are 70, 80, and 90 percent, respectively. In addition, Reliance Corporation routinely adds 10 percent of total labor and nonlabor costs to all of its projects to cover general and administrative expenses. Furthermore, Bradley adds an additional 20 percent as a contingency allowance. When he is finished, the preliminary cost estimate of the XYZ project is as follows:

Labor costs
Grade 1 $330 * 25 * 1.7$ = $14,025
Grade 2 $350 * 30 * 1.8$ = 18,900
Grade 3 $110 * 40 * 1.9$ = 8,160
                       Total labor costs   = $ 41,085.00

Nonlabor costs
Materials   = $41,500
Equipment   =   14,000
Subcontracts   =   12,000
Other   =   4,000
                       Total nonlabor costs   = $ 71,500.00
                       Subtotal   = $112,585.00

General & administrative
expenses $112,585 * 0.1   = $ 11,258.50
                       Subtotal   = $123,843.50
Contingency allowance $123,843.50 * 0.2 = $ 24,768.70
Total estimated costs $148,612.20

Table 3.2 Summary of labor hours and nonlabor costs

Work package	Labor hours by grade			Non-labor costs			
	1	2	3	Materials	Equipment	Subcontracts	Other
A	80	40	20	$ 7,000	$ 5,000		
B		30	70	$ 8,500	$ 4,000		
C	20		30				$1,000
D	100	40		$15,000	$ 1,000		
E		80				$ 5,500	
F	40	60		$ 8,000	$ 2,500		$ 500
G	90	30				$ 6,500	$ 500
H		70		$ 3,000	$ 1,500		$2,000
Total	330	350	110	$41,500	$14,000	$12,000	$4,000

3.7 THE USE OF LEARNING CURVES IN COST ESTIMATION

When estimating costs, an assumption is often made that work is completed at a constant rate. For example, in the case of an activity that is performed repeatedly, it is often assumed that the amount of time it takes to complete the activity the 20th time is not significantly different from the time it took to complete it the very first time.

To illustrate this scenario, we'll assume that a worker took 40 hours to complete an activity the first time, and that the same activity was done by the same worker three more times. Under the constant rate assumption, if the hourly labor cost is $50 per hour with an overhead rate of 50 percent, the total labor cost for this worker is as shown in Table 3.3. While this approach to cost estimation is simple and easy to use, it contains a fundamental flaw. For example, is it reasonable to assume that when repeatedly performing an activity, it will always take the same amount of time to complete? Isn't it more reasonable to assume that successive repetitions will take much less time than earlier ones? These questions form the crux of the issue of learning curve effects on project cost estimation.[11]

The learning curve shows that when an activity is performed repeatedly, there will be a successive reduction in the time required to complete that activity. In fact, many research studies on learning curves have shown that performance, in terms of time, improves by a fixed percentage each time production doubles. In other words, the labor hours per unit required to complete an activity decreases by some fixed percentage each time the output doubles.[12]

COST ESTIMATION

Table 3.3 Labor cost under a cost rate assumption

Labor rate	Activity repetitions	Overhead rate	Hours/ repetition	Total cost
$50/hour	4	1.5	40 hours	$50 * 4 * 1.5 * 40 = $12,000

For example, let's assume that it takes 80 hours of work to produce a device for the very first time. Given the learning that occurs during this initial production, the same device requires only 64 hours to complete the second time. The difference between the first and second iteration suggests a learning rate of 80 percent (64/80). This figure can be used to estimate the cost for additional units of production using the well-defined mathematical formula given below:

$$T_n = an^b$$

where
T_n = the time required for the nth unit of output
a = the time required for the initial unit of output
n = the number of units to be produced, and
b = the slope of the learning curve, expressed as log decimal learning rate/log 2

To see this formula in action, consider a project involving the manufacture of aircraft engines. This specific project requires the production of 25 units of a complex air-filtering device that the project organization has never produced.

One worker will be tasked to produce all 25 units of air filters. In addition, after production of the 20th unit, there is no significant time reduction associated with learning, and production reaches a steady state of 70 hours. We know, from previous studies, that the learning rate for workers undertaking this type of repetitive activity is 0.85. In calculating the time necessary to complete the first activity, we would apply the above values to the formula to determine the value of a, or the time needed to complete the task the first time:

$$b = \log 0.85 / \log 2$$
$$b = -0.1626 / 0.693$$
$$b = -0.235$$
$$70 \text{ hours} = a\,(20)^{-0.235}$$
$$a = 141.3 \text{ hours}$$

Note that in this example, the difference between the first and 20th unit of production represents a change in duration estimation (and therefore, cost) from over 141 hours to 70 hours for the steady state. This shows that significant differences in project cost estimates are likely, particularly when a project involves many instances of repetitive work, or large "production runs" of similar activities.

Once we have calculated the time required to produce the initial unit, we can consult tables of time multipliers to determine the appropriate multiplier for this example. Specifically, the reader should consult the portion of the learning curve cumulative values provided in Table 3.4. The multiplier is determined by identifying the value that matches 20 units of production, with a learning rate of 0.85. In this case, where the multiplier is 12.40, the time needed to produce 20 units is

	Time required for	Total time to produce
Multiplier	initial unit	20 units
(12.40) ×	141.3	= 1,752.12 hours

Because the steady state time of 70 hours occurs for the last five units, total production time required for all 25 units of this air filter is given as

$$1{,}752.12 + 70 * 5 = 2{,}102.12 \text{ hours}$$

If the labor is paid a wage rate of $50/hour with an overhead rate of 1.50, a more accurate cost estimate for this activity is:

Wage	O.H. rate	Total hours	
($50/hour) ×	(1.50)	= (2,102.12 hours)	= $157,659.00

Table 3.4 Partial table of learning curves cumulative values

	70%		75%		80%		85%	
Unit rate	Unit time	Total time	Unit time	Total time	Unit time	Total time	Unit time	Total ime
5	0.437	3.195	0.513	3.459	0.596	3.738	0.686	4.031
10	0.306	4.932	0.385	5.589	0.477	6.315	0.583	7.116
15	0.248	6.274	0.325	7.319	0.418	8.511	530	9.861
20	0.214	7.407	0.288	8.828	0.381	10.485	0.495	12.402
25	0.191	8.404	0.263	10.191	0.355	12.309	0.470	14.801
30	0.174	9.305	0.244	11.446	0.335	14.020	0.450	17.091
35	0.160	10.133	0.229	12.618	0.318	15.643	0.434	19.294
40	0.150	10.902	0.216	13.723	0.305	17.193	0.421	21.425

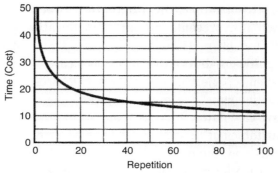

Figure 3.5 Unit learning curve log-linear model[15]
Note: Graph on arithmetic coordinates

If we had assumed a uniform production rate, the total estimated cost would have been $131,250 ($50 * 1.5 * 70 * 25). When this figure is compared to the cost of $157,659 calculated using the learning rate, we find that the assumption of uniform production understates the total production cost by $26,409 ($157,659 − 123,750). In this case, including an allowance for learning curve effects provides a much more realistic estimate of both time and cost.

Two final notes regarding the use of learning curves. First, it's possible to chart the cost of repetitive activities to ensure that cost estimates reflect the impact of learning curves. In Figure 3.5, note the curve that relates time (or cost) against activity repetition.[13] The learning curve effect phenomenon shown in this figure is typical of many projects.

Second, many books offer tables that show the total time multiplier, based on the learning rate values multiplied by the number of repetitive iterations of an activity.[14] At the same time, it is important to note that while projects with a repetitive nature lend themselves to the use of learning curves, those of a nonrepetitive nature do not. In the final analysis, project budgets should reflect the types of activities in which there is a possibility for learning curve effects to occur, and employ the appropriate mechanism to factor these effects into activity cost estimates.

REFERENCES

1. (n.d.) http://www3.bc.sympatico.ca/johnlee/dome.html; (n.d.) http://www.freep.com/news/nw/zdome21_20000721.htm; (n.d.) http://www.abc.net.au/correspondents/s97978.htm

2. (n.d.) Why IT projects falter and how astute business cases help save the day. www.loc.gov
3. Eden, C., Ackermann, F., and Williams, T. (2005) The amoebic growth of project costs. *Project Management Journal*, 36(1): 15–27.
4. Pinto, J. K. (2009) *Project Management: Achieving Competitive Advantage*. Upper Saddle River, NJ: Prentice-Hall.
5. Pinto, J. K. (2009) *Ibid.*
6. UMIST module 4, p. 4.2.5.
7. Lock, D. (2000) Managing cost. In: J. R. Turner and R. Turner (Eds.), *Gower Handbook of Project Management*, 4th ed. Aldershot, UK: Gower, pp. 293–322.
8. Pinto, J. K. (2009) *Ibid.*
9. UMIST Module 4, p. 4.2.12.
10. UMIST Module 4, p. 4.2.21.
11. Amor, J. P., and Teplitz, C. J. (1998) An efficient approximation for project composite learning curves. *Project Management Journal*, 29(3): 28–42.
12. Pinto, J. K. (2009) *Ibid.*
13. Crawford, J. R. (n.d.) *Learning Curve, Ship Curve, Rations, Related Data*. Burbank, CA: Lockheed Aircraft Corp.
14. Gray, C. F., and Larson, E. W. (2008) *Project Management: The Managerial Process*, 4th ed. Burr Ridge, IL: McGraw-Hill.
15. Crawford, J. R. (n.d.) *Ibid.*

KEY TERMS

Erroneous cost estimates
Overlooking key project costs
Low initial cost estimates
Unanticipated technical difficulties
Scope creep
Lack of or poor scope definition
Specification changes
External factors
Direct costs
Indirect costs
Recurring costs
Nonrecurring costs
Fixed costs
Variable costs
Normal costs
Expedited or crash costs
Parametric estimation
Contingency
Trend (App3)
Exponential growth (App3)
Linear trend (App3)
Nonlinear trend (App3)
Best fit (App3)
Least squares regression (App3)
Auto catalytic growth function (App3)

Appendix to Chapter 3

Forecasting Methods for Cost and Value Management

This appendix is not for everyone. The science of forecasting is complex, and, at times, employs some strenuous mathematics. So...why is it here?

As we completed Chapter 3, we felt that some discussion of forecasting methods was warranted. We recognize that readers could readily embrace the material in this chapter without going deeply into various forecasting methods, but there are likely to be readers who want this additional information. If you are among the latter, we invite you to read on—the forecasting methods we present will be of great help in preparing reasonable cost estimates for your projects.

Forecasting takes place at the front end of the project environment. It is a critical, fundamental process for any business, and project organizations are no exceptions. Forecasting demand, time required for activities, and costs associated with those activities creates critical input for project planning and for controlling project progress. Forecasting is also one of the primary means for tracking progress or for changing direction if changes in the project environment warrant it.

In this appendix, you will learn about the types of forecasts needed to effectively manage the cost and value of projects, along with some important techniques that can be used to generate them.

3A.1   CATEGORIES OF FORECASTING IN PROJECT MANAGEMENT

The ability to forecast what can occur is a vital aspect of project management, and can be broadly classified in three categories:

- Forecasting individual events

68 FORECASTING METHODS FOR COST AND VALUE MANAGEMENT

- Forecasting the outcome or progress of either a process or a large group of activities
- Forecasting technology

Although all three categories are somewhat different in character, some forecasting methods can be used across all three.

Of the first two categories, individual events can be forecast in a typical project situation, where the completion dates of individual activities are estimated as the project progresses. Forecasting the outcome of the project as a whole is a different process, because it requires that the project's past history be created, assuming that relevant data from the past has been collected and retained. This history can serve as the basis for generating subsequent forecasts to track progress. Forecasts of individual, stand-alone events that are not the result of a string of events are generally forecast by opinion-based methods, whereas events that are the result of a process can be forecast by one of a number of trend extrapolation techniques.[1] The final category—technology forecasting—is another vital, ongoing activity in the project management discipline, because its rapid evolution can render a current project facility obsolete.

Within this basic division, two further distinctions in forecasting can be made:

- Where there is some measurable history to use as a basis for the forecast.
- Where there is little in the way of history.

3A.2 FORECASTING METHODS FOR PROJECTS

Forecasting methods for projects fall into two broad categories: qualitative and quantitative. Qualitative methods are used where no reliable, historical, or statistical data are available. Quantitative techniques, on the other hand, are appropriate in project situations where measurable, historical data are available and are usually used in forecasting for short or intermediate timeframes. These techniques can be classified into two broad categories: time series analysis and causal methods.

3A.3 TIME SERIES ANALYSIS

The main theme underlying time series analysis is that past behavior of data can be used to predict future behavior. As such, a time series is defined as a sequence of observations taken at regular intervals over a period of time. For example, a time series for a project organization such as Rolls Royce would consist of the monthly demand data for a particular type of aircraft engine over the past ten years. Analyzing time series data involves identifying and evaluating its major components, including trend, seasonality, cycle, and random. In this discussion, we focus our attention on the trend component of a time series, as it is the most relevant component in a project environment.

Trend is the long-term movement of data over time. This definition implies that time is the independent variable, and the data or set of observations we are interested in is the dependent variable. When we track data purely as a function of time, there are several possible scenarios. First, data may exhibit **no trend**. In this case, the set of data remains constant and is unaffected by time. The second possible scenario is **linear trend**, where data as a function of time have a linear relationship. It should be noted that data can exhibit a positive linear trend with a rate increase between successive periods, or a negative linear trend with a rate of decrease between successive observations.

The behavior of data over time can also exhibit a **nonlinear trend** pattern, such as **exponential growth or decay**. In the case of exponential growth, each succeeding observation increases by some constant factor, whereas in exponential decay, each succeeding observation decreases by some constant factor.

3A.4 LINEAR REGRESSION ANALYSIS

One of the most useful techniques to evaluate and forecast the trend component of a time series is regression analysis. The trend component may or may not be linear. For example, we saw in a previous chapter that the typical S-curves we use in managing projects usually exhibit a nonlinear trend. However, there are several project situations where the relationship of a project variable as a function of time *can* be assumed to have a linear trend, such as forecasting the cost of an activity as a function of time.

70 FORECASTING METHODS FOR COST AND VALUE MANAGEMENT

For example, due to wage and materials increases, inflation, and other factors, it takes more money today to build a house than it did 20 years ago.

While the assumption of linearity may not hold true in many situations over the long term, a linear relationship is often a reasonable assumption in the intermediate timeframe. In these cases, linear regression analysis is a viable forecasting technique. It is also a useful methodology for determining the empirical relationship between two project variables if the underlying reasons for the relationship between those variables can be hypothesized to be approximately linear.

To evaluate the trend component of a time series, we use linear regression analysis to develop a linear trend equation. This is the equation of a **straight line** as given by

$$y_t = a + bt + e$$

where b is the slope (gradient) of the line and a is the intercept with the y axis at $t = 0$, y_t is the value of the project variable (for which we desire a forecast) at time period t, and e is the forecast error.

The technique of linear regression analysis involves determining the values for a and b for a given data set. From a graphical perspective, the technique involves drawing a straight line that best fits the scatter of observed values that have been plotted over time (see Figure 3A.1). **Best fit** means that the differences between the actual y values and predicted y values are at a minimum. However, because positive differences will offset negative differences, they are squared.

Mathematically, the best-fitting line is the one in which the sum of the squares of the deviations of all the data points from the calculated line is a minimum. In essence, we are choosing a line where the scatter of the observed data about the line is at its smallest. The technique of

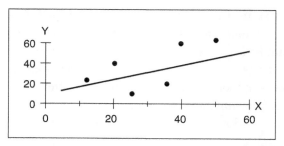

Figure 3A.1   Line of best fit in regression analysis

least squares regression minimizes this sum of the squared differences or errors. By using this technique, we can get formulas for a (the intercept) and b (the slope of the regression line). These formulas are given below for a time series of n points of data, where t = time, i.e., the number of time periods from the starting point, and y = the observed value in a given time period. The slope b of the line is given by

$$b = \frac{n \Sigma t y - \Sigma t \Sigma y}{n \Sigma t^2 - (\Sigma t)^2}$$

The intercept a is given by

$$a = \frac{\Sigma y - b \Sigma t}{n}$$

Interpretation of coefficients
- Slope b: the estimated y changes by b for each one unit increase in t.
- y intercept (a): the average value of y when $t = 0$.

Let's look at a simple example for developing a linear trend equation using linear regression analysis for forecasting the trend component of a time series.

Example of Linear Regression

We know the cost of a project activity, based on past data, and wish to estimate the probable cost for a project that starts next month. We could simply plug in old numbers, but they may be out of date, which would lead to serious over- or underestimation of costs. However, we do have some historical information on how much this activity has cost in past jobs. The historical data on the cost (in hundreds of $) for the project activity is given in Table 3A.1. With this in hand, we can develop a trend equation using linear regression analysis and forecast the cost of this activity for periods 10 and 15 of our project. The calculations for determining the slope and intercept of the regression line are shown in Table 3A.2. The slope b of the line is given by

$$b = \frac{n \Sigma t y - \Sigma t \Sigma y}{n \Sigma t^2 - (\Sigma t)^2} = \frac{9 * 3,084 - 45 * 592}{9 * 285 - (45)^2} = 2.067$$

Table 3A.1

T = year	Y = activity cost (in $ hundreds)
1	58
2	57
3	61
4	64
5	67
6	71
7	71
8	72
9	71

Table 3A.2

t	y	$t*y$	t^2	y^2
1	58	58	1	3,364
2	57	114	4	3,249
3	61	183	9	3,721
4	64	256	16	4,096
5	67	335	25	4,489
6	71	426	36	5,041
7	71	497	49	5,041
8	72	576	64	5,184
9	71	639	81	5,041
$\sum t = 45$	$\sum y = 592$	$\sum t*y = 3,084$	$\sum t^2 = 285$	$\sum y^2 = 39,226$

The intercept a of the line is given by

$$a = \frac{\sum y - b \sum t}{n} = \frac{592 - 2.067 * 45}{9} = 55.44$$

The resulting linear trend equation is given by

$$y_t = a + bt = 55.44 + 2.067t$$

The forecast for period 10 is given by

$$y_{10} = 55.44 + 2.067 * 10 = 76.11$$

and for $t = 15$,

$$y_{15} = 55.44 + 2.067 * 15 = 86.445$$

Based on linear trends using regression analysis, we can now estimate that the activity will cost $7,611 for period 10 and $8,644.50 for period 15.

In the discussion on linear regression analysis and the example above, the independent variable was t (the time period). However, the linear regression technique can also be used to determine association or causation between two variables; that is, the direct relationship or impact of one factor on another (for example, the manner in which smoking increases one's risk of heart disease). In such cases, we use the notation x for the independent variable and y for the dependent variable. The linear regression equation in such cases would be

$$y_c = a + bx_i$$

where the slope b of the regression line is given by

$$b = \frac{n \sum x_i y_i - \sum x_i \sum y_i}{n \sum x_i^2 - (\sum x_i)^2}$$

and the intercept a of the regression line is given by

$$a = \frac{\sum y_i - b \sum x_i}{n}$$

3A.4.1 Evaluating the "Fit" of the Regression Line

After obtaining the regression line, the next step in regression analysis is to evaluate how well the model describes the relationship between variables, or how good the line of best fit is. Three measures can be used to evaluate how well the computed regression line fits the data: the coefficient of determination (R^2), the correlation coefficient (r), and the standard error of the estimate (s_{yx}).

1. *The coefficient of determination (R^2)*—Four measures of variation can be computed in linear regression analysis:

    - *Total sum of squares (SST)*—Measures the variation of the actual y values around the mean Y:

    $$\text{SST} = \sum (y_i - \bar{y})^2$$

- *Explained variation (SSR) or the regression sum of squares*—Measures the variation due to the relationship between x and y, i.e., the difference between the mean y and the predicted value y using regression:

$$\text{SSR} = \Sigma(\hat{y}_i - \bar{y})^2$$

- *Unexplained variation (SSE) or error sum of squares*—Measures variation not explained by regression, i.e., variation due to other factors or variables not included in the regression model:

$$\text{SSE} = \Sigma(y_i - \hat{y}_i)^2$$

- *Coefficient of Determination (R^2)*—The proportion of variation explained by regression, i.e., by the relationship between x and y. The formula used to compute R^2 is

$$R^2 = \frac{\text{SSR}}{\text{SST}} = \frac{\Sigma(\hat{y}_i - \bar{y})^2}{\Sigma(y_i - \bar{y})^2} = \frac{a\Sigma y_i + b\Sigma x_i y_i - n(\bar{y})^2}{\Sigma y_i^2 - n(\bar{y})^2}$$

R^2 takes on a value between 0 and 1. The higher the value of R^2, the better the line of fit. In other words, the higher the R^2 value, the more confidence we can have that the estimate is accurate.

2. *The correlation coefficient (r)*—This statistic measures the strength of the relationship, or association between x and y, and takes on a value between -1 and $+1$. The closer the value of r to $+1$ or -1, the stronger the relationship between the variables. The formula for r is given by

$$r = \frac{\Sigma x_i y_i - n\bar{x}\bar{y}}{\sqrt{\Sigma x_i^2 - n(\bar{x})^2}\sqrt{\Sigma y_i^2 - n(\bar{y})^2}}$$

3. *The standard error of the estimate (s_{yx})*—Random variation, which is the variation of the actual (observed) y values from the predicted y values ($\hat{y}_i$), is measured by the standard error of the estimate. The smaller the value of s_{yx}, the smaller the variation of the actual (observed) y values from the predicted y values ($\hat{y}_i$), and the better the fit between variables. The interpretation of the standard error of the estimate is similar to that of standard deviation, which measures the variability or dispersion around the arithmetic mean.

In other words, the standard error of the estimate measures the variability of the actual y values around the fitted regression line:

$$S_{yx} = \frac{\sqrt{\Sigma(y_i - \hat{y}_i)^2}}{\sqrt{(n-2)}} = \frac{\sqrt{\Sigma y_i^2 - a\Sigma y_i - b\Sigma x_i y_i}}{\sqrt{(n-2)}}$$

Let's now compute the values for R^2, r, and s_{yx} for the example problem on linear regression that we solved earlier. Because the independent variable in this example is t instead of x, we solve the above formulas for R^2, r, and s_{yx} by substituting t for x. We have the following data from the example above:

$$n = 9,\ \Sigma t = 45,\ \Sigma y = 592,\ \Sigma t*y = 3,084,$$
$$\Sigma t^2 = 285,\ \Sigma y^2 = 39,226,\ a = 55.44, b = 2.067$$

$$R^2 = \frac{a\Sigma y_i + b\Sigma t_i y_i - n(\bar{y})^2}{\Sigma y_i^2 - n(\bar{y})^2}$$

$$= \frac{(55.44*592) + (2.067*3084) - 9*(592/9)^2}{39,226 - 9*(592/9)^2} = 0.897$$

$$r = \frac{\Sigma t_i y_i - n\bar{t}\bar{y}}{\sqrt{\Sigma t_i^2 - n(\bar{t})^2}\sqrt{\Sigma y_i^2 - n(\bar{y})^2}}$$

$$= \frac{3,084 - 9*(45/9)*(592/9)}{\sqrt{285 - 9*(45/9)^2}\sqrt{39226 - 9*(592/9)^2}} = 0.947$$

$$S_{yt} = \frac{\sqrt{\Sigma y_i^2 - a\Sigma y_i - b\Sigma t_i y_i}}{\sqrt{(n-2)}}$$

$$= \frac{\sqrt{39226 - 55.44*592 - 2.067*3084}}{\sqrt{(9-2)}} = 2.046$$

All of the statistics indicate that the regression line obtained for example 1 is a very good line of fit. Put another way, this means that our regression exercise gives us the ability to predict activity costs for this example with a high degree of confidence. Specifically, the value of R^2 of 0.897 means that 89.7 percent of the variation in the y values is accounted for by changes in the t value. Similarly, the r value of 0.947 indicates that the variables t and y are almost perfectly positively correlated (close to the value of 1). Finally, the S_{yt} value of 2.046 indicates that actual y values fluctuate in relation to the predicted y values, represented by the regression line, by about 2.46 units.

In the example shown above, only two variables were considered. However, there may be occasions where several independent variables have an influence on the dependent variables. In such cases, an extension of linear regression technique called **multiple linear regression** can be used. A multiple linear regression equation is

$$y = b_0 + b_1 x_1 + b_2 x_2 + b_3 x_3 + \cdots + b_i x_i + e,$$

where x_1, x_2, x_3, and x_i are independent variables, and e is the error term; b_0, b_1, b_2, b_3, and b_i are termed the regression coefficients, and represent the amount by which y changes for one increment of x_i, assuming all other independent variables are held constant.

A brief illustration of the multiple regression technique was presented earlier in Chapter 3, in the context of parametric estimation. Readers who are interested in learning more about this procedure should consult a statistics text.

There are, of course, many occasions when a simple linear model will not adequately describe the relationship between one variable and another. Some curvilinear relationships, such as the learning curves frequently encountered in engineering project work, can be described by the general expression:

$$Y = kx^n$$

The above expression can be converted into logarithmic plots. These are an extremely useful way of analyzing trends, because many phenomena progress as a simple law of the form $y = kx^n$. The rate of progress of the performance of some technologies can be clearly illustrated by linear logarithmic trends over time, and some frequency distributions can be illustrated using logarithmic plots. Once a straight line has been established, it is a simple process to compute the underlying equation. In addition, a straight line can be used as the basis for forecasting future values of the observed variable, simply by projecting it forward. However, we must be careful about just how far ahead we can project to generate forecasts.[2]

3A.5 S-CURVE FORECASTING

Another example of a curvilinear relationship is S-curve behavior in projects. The typical form of the relationship between project duration and expenditures incurred is S-shaped, where budget expenditures are initially low and increase rapidly during the major project execution stage before starting to level off again as the project gets nearer to completion.

The S-curve figure represents the project budget baseline against which actual cumulative budget expenditures will be evaluated. It helps project managers understand the correlation between project duration and budget expenditures, and provides a good sense of where the highest levels of budget spending are likely to occur. Forecasting the S-curve can also help project managers generate estimates of expenditures during various stages of project duration.

The S-curve (also called a pearl curve) is based on what is known as the logistic or **auto catalytic growth function**. The URL below contains a complete discussion on the theory and steps involved in S-curve forecasting. We strongly urge readers to go over this document and follow through on the worked-out example.

http://leeds-faculty.colorado.edu/Lawrence/Tools?SCurve/scurve.xls

Below is an exercise that illustrates the S-curve forecasting methodology using the procedure presented at the above website. The data provided in Table 3A.3 show cumulative percentages of work actually completed for a construction project.

Table 3A.3

Month	Cumulative percentage of work completed
1	4.7
2	7.1
3	9.5
4	12.5
5	17.0
6	22.0
7	29.0
8	36.0
9	46.0
10	56.0
11	64.0
12	72.0

a. Graph the data in Excel and comment on the shape of the distribution.
b. Forecast the cumulative percentages of work completed for each of the next six months' linear regression.
c. Is linear regression an appropriate forecasting method for these data? If not, can you suggest an alternative forecasting method?
d. Use this method to forecast the percentages of work completed for months 19, 20, 21, and 22.
e. Is this a better forecasting method for these data than linear regression? How do you know?

Solutions

a. *Plot of the data (Figure 3A.2)*—The shape of the distribution suggests that the project progresses at a slow pace through period 4. The per-month completion rate of work begins to accelerate in period 5, and continues to increase through period 10. The progress begins to level off beginning from period 11, and continues that trend in period 12. The cumulative percentage of work completed exhibits typical S-curve behavior.

b. *Linear regression solution (Table 3A.4)* The slope b of the line is given by

$$b = \frac{n\Sigma ty - \Sigma t \Sigma y}{n\Sigma t^2 - (\Sigma t)^2} = \frac{12 * 3347.4 - 78 * 375.8}{12 * 650 - (78)^2} = 6.327$$

The intercept a of the line is given by

$$a = \frac{\Sigma y - b\Sigma t}{n} = \frac{375.8 - 6.327 * 78}{12} = -9.809$$

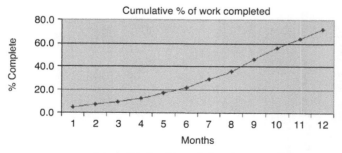

Figure 3A.2 Plot of the data

Table 3A.4

Month (t)	Cumulative % of work completed (y)	t * y	t * t	y * y
1	4.7	4.7	1	22.09
2	7.1	14.2	4	50.41
3	9.5	28.5	9	90.25
4	12.5	50.0	16	156.25
5	17.0	85.0	25	289.00
6	22.0	132.0	36	484.00
7	29.0	203.0	49	841.00
8	36.0	288.0	64	1,296.00
9	46.0	414.0	81	2,116.00
10	56.0	560.0	100	3,136.00
11	64.0	704.0	121	4,096.00
12	72.0	864.0	144	5,184.00
$\Sigma_t = 78$	$\Sigma_y = 375.8$	$\Sigma_{ty} = 3347.4$	$\Sigma_t 2 = 650$	$\Sigma_y 2 = 17,761.00$

The resulting linear trend equation is given by

$$y_t = a + bt = -9.809 + 6.327t$$

The forecast for periods 13 to 18 is given by

$$y_{13} = -9.809 + 6.327 * 13 = 72.44$$
$$y_{14} = -9.809 + 6.327 * 14 = 78.77$$
$$y_{15} = -9.809 + 6.327 * 15 = 85.10$$
$$y_{16} = -9.809 + 6.327 * 16 = 91.42$$
$$y_{17} = -9.809 + 6.327 * 17 = 97.75$$
$$y_{18} = -9.809 + 6.327 * 18 = 104.08$$

c. It appears that linear regression is not an appropriate forecasting method for these data. The solutions to parts a and b clearly indicate that data exhibit an S-curve behavior. As a result, this problem should be solved using the S-curve forecasting methodology.

d. *S-curve forecasting*—The solution to the above example using the S-curve forecasting methodology is given in the table in Figure 3A.3. The column titled "Period" (**2**) refers to the number of periods of data. Column (**4**) contains the data for the cumulative percentage of

80 FORECASTING METHODS FOR COST AND VALUE MANAGEMENT

(1) Data Title: Question 5: S-curve for forecasting (%work completed)

					Logistic Function			Exponential Function		Linear Function	
			Current Max:	72.72	α = 3.3512 β = −0.3553		a = 28.54 b = 0.36				
						(8) L =	100				
			(4) y			SD Errors = 47.2		SD Errors = 310.6		SD Errors = 46.2	
(2) t	(3) Label		Cumulative % of	(5)		(6a)	(7a)	(6b)	(7b)	(6c)	(7c)
Period	Months		work completed	Y	Transform	Forecast	Errors ε	Forecast y'	Errors ε	Forecast y'	Errors ε
1	January		4.7		3.0095	4.8	0.1	5.8	1	5.1	0
2	February		7.1		2.5714	6.7	−0.4	7.4	0	10.3	3
3	March		9.5		2.2541	9.2	−0.3	9.5	0	15.4	6
4	April		12.5		1.9459	12.7	0.2	12.2	0	20.6	8
5	May		17.0		1.5856	17.2	0.2	15.6	−1	25.7	9
6	June		22.0		1.2657	22.8	0.8	20.0	−2	30.9	9
7	July		29.0		0.8954	29.6	0.6	25.7	−3	36.0	7
8	August		36.0		0.5754	37.5	1.5	32.9	−3	41.2	5
9	September		46.0		0.1603	46.2	0.2	42.3	−4	46.3	0
10	October		56.0		−0.2412	55.0	−1.0	54.2	−2	51.5	−5
11	November		64.0		−0.5754	63.6	−0.4	69.6	6	56.6	−7
12	December		72.0		−0.9445	71.3	−0.7	89.2	17	61.8	−10
13	January				#DIV/0!	78.0	78.0	114.5	114	66.9	67
14	February				#DIV/0!	83.5	83.5	146.9	147	72.1	72
15	March				#DIV/0!	87.8	87.8	188.5	188	77.2	77
16	April				#DIV/0!	91.2	91.2	241.8	242	82.4	82
17	May				#DIV/0!	93.6	93.6	310.2	310	87.5	88
18	June				#DIV/0!	95.5	95.5	398.0	398	92.7	93
19	July				#DIV/0!	96.8	96.8	510.7	511	97.8	98
20	August				#DIV/0!	97.7	97.7	655.2	655	103.0	103
21	September				#DIV/0!	98.4	98.4	840.6	841	108.1	108
22	October				#DIV/0!	98.9	98.9	1,078.4	1,078	113.3	113

Figure 3A.3   S-curve forecasting results

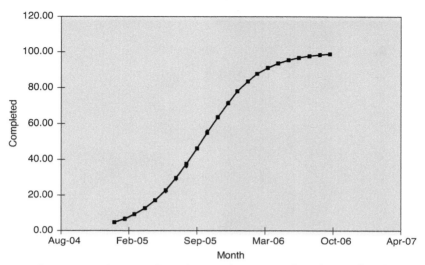

Figure 3A.4 S-curve: Cumulative percentage of work completed

work completed. Column (**6a**) under the heading "Logistic Function" contains the forecasts generated using the S-curve methodology. Consequently, the forecast of the "cumulative percentage of work completed" for the next six periods (19 to 22) are 96.8, 97.7, 98.4, 98.9, respectively. Figure 3A.4 shows the logistics plot of S-curve forecasts.

e. The S-curve forecast is a better forecasting method for this data, as opposed to the linear regression method. For this problem, linear regression will indicate a continuously rising trend, regardless of the pattern developed by the actual data. Clearly, data presented in this problem for the first 12 periods are beginning to assume the shape of an S-curve. This is the pattern often seen in projects, in that work tapers off at the end as the project enters the delivery phase and resources are reassigned. Also, linear regression would indicate that the project should be completed shortly after period 17. As the S-curve forecasts indicate, this is clearly not the case.

REFERENCES

1. UMIST Module 4, Chapter 8, p. 4.8.1.
2. UMIST Module 4, p. 4.8.24.

Chapter 4

Project Budgeting

LEARNING OBJECTIVES

- Describe the issues in project budgeting.
- Define activity-based costing.
- Discuss the advantages and disadvantages of top-down and bottom-up budgeting.
- Describe the effects of crashing a project on the project budget.

In the last chapter, we examined the various concepts, approaches, and problems related to cost estimation. Here, we discuss project budgeting, which is inextricably linked with estimation in that both processes deal with the cost of completing an activity, work package, or project. Once cost estimates are approved, they become project budgets that allocate resources with an agreed-upon, contracted amount of what work should cost.

Budgets serve a vital role in the management of projects. They function as a control mechanism that sets the standard against which future expenditures will be monitored. With timely data collection and reporting, they enable the project team to identify and report current problems, and to anticipate future ones. And when done properly, they relate the use of resources with the achievement of corporate goals.

In most organizations, there are two kinds of budgets: **project budgets** and **fiscal operating budgets**. The difference between the two is that a project budget covers the entire duration of a single project, while a fiscal operating budget applies to a single year. In this chapter, we examine the various aspects and approaches to project budgeting, including time-phased project budgeting.

4.1 ISSUES IN PROJECT BUDGETING

A project budget identifies the project's allocated resources, goals, and the schedule that allows the organization to achieve those goals. Developing a project budget requires not only a knowledge of the resources needed, but also information regarding how many will be needed and when, as well as how much they will cost.

Because each project is unique, the budgeting process for project organizations is considerably more complex than for traditional organizations. Project managers may not have the luxury of accessing and using the historical information that is typically available for ongoing, routine activities. And, while there may be budgets and reports from similar projects undertaken in the past, these can only serve as rough guides. Consequently, all project budgets are based on estimates of resource usage and their associated costs.

The budgeting process gets even more complicated for multiyear projects, because plans, schedules, resources, and costs are set early in the project life cycle and may change in future years. This can happen due to a number of factors, including the availability of newer equipment, more skilled personnel, and alternative materials for completing project activities.

Another budgetary issue for project managers is the way in which resources are allocated over a project's different time periods. In many cases, actual usage of resources may be very different from the accounting department's assumptions. Assume, for example, that $10,000 of a given resource will be used to complete an activity over a five-week period. The actual usage of this resource is $3,000 during week one, none in week two, $2,500 in week three, $4,000 in week four, and $500 in week five. If this information about the pattern of resource usage is not included in the plan, the accounting department may distribute the expenditures equally over the five weeks. While this situation will have no impact on the overall project budget, it certainly will have an impact on the timing of cash flow.

Another issue is that accounting systems and practices typically vary from one project organization to the next. The project manager must have a thorough understanding of, and familiarity with, the systems and practices in his or her organization, or it will be impossible to exercise budgetary control.

Finally, in preparing a project budget, the project manager will need to ensure that each expenditure is identified and tied to a specific project activity and its associated milestone. The mechanism through which the project manager can accomplish this is the WBS, which has a unique account number that can be charged as and when work is completed for each element. This aspect of project budgeting is absolutely vital.

4.2 DEVELOPING A PROJECT BUDGET

Essentially, the project budget is a plan that incorporates the allocation of resources to various work packages and departments, along with a schedule to ensure that the company is in a position to achieve project goals. The two must be developed concurrently, because the budget will provide a clear picture of whether or not project milestones can be achieved.

Developing a project budget involves estimation of costs, subsequent analyses, frequent revisions, and, to a certain extent, intuition. Meaningful budgets are developed through frequent interaction among concerned parties, and require data input from a variety of sources. Regardless of the type of project budget, the most important caveat is that it should be in alignment with and reflect support for both the project and overall corporate goals.

4.2.1 Issues in Creating a Project Budget

A number of important issues go into the creation of the project budget, including the process by which the project team and organization gather data for cost estimates, budget projections, cash flow income and expenses, and so forth. The methods for data gathering and allocation can vary widely across organizations. For example, while it may be common to allocate an amount to cover the cost of producing one work package in the overall project, some firms will allocate that amount as a one-time expense, while others will recognize that the costs accrue over the period of time necessary to complete the work package.

While the method used may not matter to the project's overall budget, the difference can have a significant impact on cash flow. For this reason, it is important to recognize the accounting and budgeting system used within a particular organization when preparing the project's budget. This will minimize the conflicts that can arise with cost control personnel when budget data are misinterpreted.

4.3 APPROACHES TO DEVELOPING A PROJECT BUDGET

The way in which cost data are collected and interpreted is determined by whether the project organization takes a top-down or bottom-up approach to budgeting. The two approaches use radically different methods for collecting relevant information, which can lead to completely different results. Let's take a look at both approaches.

4.3.1 Top-down Budgeting

The top-down approach utilizes the judgment and experience of top and senior-level managers, as well as past data on similar activities. It assumes that experience in past projects will enable senior management to provide guidance on accurate cost estimates for future projects.

The rationale behind top-down budgeting is that top management may not have the inclination or the time to estimate the precise cost of each and every element of a project. What they do have, however, is the experience, knowledge, and savvy to provide a reasonable aggregate estimate of the overall budget. To provide more detailed estimates for the various work packages and their individual tasks, personnel in the lower levels of the organization (where the work packages and tasks will be accomplished) must be involved.

Here's how the top-down approach works in practice. Senior managers first estimate the overall costs of the project, as well as the costs of major deliverables. These projections are then handed down to lower level managers, who further disaggregate the projections into detailed budget estimates for the specific work packages that comprise the deliverables. This process is continued at each step down the hierarchy, until the project is broken into and ultimately evaluated on a task-by-task basis at the lowest level of the project organization's hierarchy (see Figure 4.1).

Top-down Budgeting: Advantages

The advantage of **top-down budgeting** is that top management's estimates of project costs, in aggregate terms, often tend to be quite accurate. Furthermore, these aggregate estimates provide the basis for disaggregation, which in turn provides important budgetary discipline and cost control. Also, because senior-level managers are directly involved in developing the project budget, there is a sense of commitment, ownership, and support.

Approaches to Developing a Project Budget

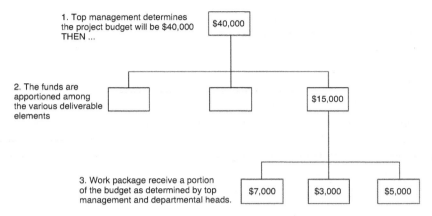

Figure 4.1 Top-down budgeting

Top-down Budgeting: Disadvantages

When used correctly, top-down budgeting can serve as a viable method for cost allocation, but it must be used in accordance with the philosophy that underlies it. That is, we assume that top management may not have the inclination or the time to estimate the precise cost of each and every element of the project. However, it is expected that they do have the experience, the knowledge, and the savvy to provide a reasonable aggregate estimate of the overall project budget. If this rationale is incorrect it can result in some clear disadvantages when top-down budgeting is employed.

First, lower-level managers may feel that their share of the project budget doesn't accurately reflect the tasks to be accomplished. Should senior managers remain steadfast in their budgetary position, lower-level managers may feel that they have no choice other than accepting what they perceive as insufficient budgetary allocations. As a result, they may feel that they bear all the responsibility for their part of the project, should it fail, but none of the authority to recognize and support a reasonable budget for their activities.

Second, top-down budgeting creates a potentially conflict-laden atmosphere as the lower levels of the organization work to align their specific task budgets with the overall project budget. Because various departments are competing for a share of a fixed amount, the top-down approach can have the net effect of pitting one department against another and creating adversarial relationships. When functional managers view the budgeting process in this competitive light, there is a strong potential for political behavior, inflated estimates, and other gamesmanship to justify inaccurate or deliberately excessive budget requests.

Third, the effectiveness of top-down budgeting depends on the accuracy and honesty with which it is used. The technique is truly useful only when it is coupled with a top management that is knowledgeable enough to make reasonable cost estimates of the project at hand; if this is not the case, the project is immediately saddled with an unrealistic budget that cannot be achieved.

Finally, some executives use top-down budgeting inappropriately, choosing to view it as a motivational technique for lower-level managers. The cost of a project can also be deliberately underestimated in the belief that it will spur cost savings and efficiencies from subordinates. Either of these approaches is more likely to result in staff demoralization and budget overruns than project success.

4.3.2 Bottom-up Budgeting

The most important prerequisite for the bottom-up approach is the availability of a detailed WBS that identifies all elements in the project, e.g., deliverables, subdeliverables, and work packages. Utilizing the framework provided by the WBS, the bottom-up approach begins with the construction of individual budgets that assign both direct and indirect costs associated with labor, materials, and overhead for the most elemental tasks. An important part of this process is consulting the personnel who regularly perform these tasks—gathering detailed information regarding time and budget from those actually responsible ensures highly accurate results. The resulting task-level costs are then aggregated to create budgets at the deliverable level. Subsequently, these deliverable budgets are aggregated to create the overall project budget. An example of the bottom-up budgeting process is shown in Figure 4.2.

While it is critical that as many task elements be identified and included in a bottom-up budget as possible, it is quite difficult to develop a complete list, particularly during the early stages of the project life cycle. The finished, aggregated budget is then presented to top managers, who merge and streamline the various budget proposals and ensure that there is no overlap or double counting of requested resources. No matter

Approaches to Developing a Project Budget   89

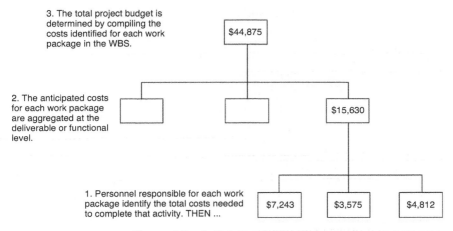

Figure 4.2   Bottom-up budgeting

where the budget originates, top management is ultimately responsible for creating the organization's final master budget.

Bottom-up Budgeting: Advantages

The advantage of bottom-up budgeting, first and foremost, is that it forces the creation of detailed project plans as an initial step for creating a budget. Second, the approach facilitates participative management, because it assumes that the individuals actually performing the tasks are more likely to have a clear understanding of resource requirements than their superiors. As a result, even the lowest levels of the organization become involved in the project budgeting process, which increases its chances of acceptance. In addition, this approach provides junior-level managers with invaluable training, experience, and knowledge in developing a project budget.

The third advantage of bottom-up budgeting is that it facilitates coordination between project managers and the functional department managers who must consider various resource requests as they develop their own budgets. Finally, because bottom-up budgeting emphasizes the unique creation of budgets for each project, it allows top management to prioritize among projects that are competing for resources. If it is determined that there are simply more projects than money available, the process of building a budget from the ground up gives top management additional information to use in determining which projects they are willing to support.

Bottom-up Budgeting: Disadvantages

The major disadvantage of bottom-up budgeting is the reduced role played by top management in the initiation and control of the budgeting process. In other words, the impetus for developing project budgets must now come from lower level managers, while top management's role is limited to analyzing individual budgets presented to them. This reduction in top management's role has the potential to create a significant division between the organization's strategic and operational activities. For this reason, the bottom-up budgeting approach is less popular within many organizations.

If we accept that the budget is the most important control tool for the project organization, it is understandable that senior-level managers are reluctant to relinquish control to inexperienced junior-level managers and personnel. In addition, there is the possibility of budgetary game-playing by junior-level managers who overstate their budget requirements. From a practical, self-protection perspective, there is huge incentive for junior project team personnel to exaggerate the costs of their activities to give themselves "wiggle room." Finally, the bottom-up approach can also be a time-consuming process for top management due to the repetitive fine-tuning, adjustments, and delays associated with budget resubmission.

4.4 ACTIVITY-BASED COSTING

Activity-based costing (ABC) is frequently used for project budgeting. The basis of activity-based costing is that projects consume activities, and activities consume resources. As such, costs are initially assigned to activities (the discrete tasks that need to be completed to deliver the project), and then assigned to projects, based on each project's use of resources.

4.4.1 Steps in Activity-based Costing

Activity-based costing consists of four steps[1]:

1. It begins with the identification of activities that make use of resources, and assigns costs to them. This step involves using the WBS to break the project down into its work packages. Costs are then assigned to each work package based on the resources required to complete each identified activity.

2. The cost drivers associated with the work package are identified. These are elements that cause, or "drive," an activity's costs. For example, the principle cost driver for many project activities is human resources in the form of labor. Similarly, an important cost driver for construction projects is the variety of raw and finished materials needed.
3. Next, a cost rate per unit of the cost driver is calculated. For example, human resource values can be stated as cost of labor per hour.
4. To assign costs to a project that utilizes a cost driver, the cost driver rate per unit is multiplied by the total number of cost driver units consumed. For example, if the cost rate for a designer is $50/hour, and 100 hours of this designer's time is used on the project, the cost assigned to the project for this designer is

$$\$50/\text{hour} * 100 \text{ hours} = \$5,000.00$$

4.4.2 Cost Drivers in Activity-based Costing

In a typical project, there are several cost drivers relating to both direct and indirect project costs, such as overhead expenses. It is imperative to identify principle cost drivers early in the project-planning phase. This will ensure that budget projections are based on accurate information regarding the types of expenditures that will be incurred, the cost rates to be charged, and the number of cost driver units to be used. The viability and effectiveness of activity-based costing hinges on this early identification of the principle cost drivers and the subsequent creation of a meaningful control document.

4.4.3 Sample Project Budget 1

Table 4.1 illustrates a sample portion of a project budget. In this table, both direct and indirect costs are identified for each project activity, and budget lines are assigned. The goal of this preliminary budget is to identify direct costs and indirect costs such as overhead expenses. At times, it may become necessary to break the overhead expenses to finer levels of detail, such as expenses relating to health insurance. However, this is usually more appropriate when a detailed project budget is developed.

Table 4.1 Sample portion of a project budget

	Budget		
Activity	Direct costs	Overhead	Total cost
Design	4,200	800	5,000
Blueprint	6,900	1,100	8,000
Engineering	4,500	1,000	5,500
Test	5,250	1,250	6,500
Sequencing	9,000	3,000	12,000
Production analysis	3,500	1,500	5,000

4.4.4 Sample Project Budget 2

The second sample budget, shown in Table 4.2, takes the total planned expenses identified in Table 4.1 and compares these figures against actual accrued project expenses. This budget can be used for variance reporting through periodic updating to show differences, both positive and negative, between the baseline budget assigned to each activity and the actual cost of completing those tasks.

The benefits of this budget are that it offers a central location for compiling all relevant project cost data, and provides comparisons between planned and actual budget expenditures. It also allows for the preliminary development of variance reports. The disadvantage of using such a static budget document is that it does not reflect the project schedule or the fact that activities are phased in over time. Because of this important limitation, a variance-to-actual budget, as shown in Table 4.2, may not be as effective a

Table 4.2 Sample budget showing planned and actual activity costs and variances

	Budget		
Activity	Planned	Actual	Variance
Design	5,000	4,250	750
Blueprint	8,000	8,000	0
Engineering	5,500	3,500	2,000
Test	6,500	8,500	(2500)
Sequencing	12,000	11,250	750
Production analysis	5,000	5,150	(150)
Total	42,000	40,650	850

control device for the project as a time-phased budget that allows the project team to evaluate budgets in relation to the schedule baseline.

4.5 PROGRAM BUDGETING

The proliferation of projects in many firms has created a serious need to organize budgets in a manner that accurately tracks and controls project expenditures. With traditional budgeting methods, the budget for a project may be spread over different organizational units; for example, assigning a portion of the costs to marketing, engineering, production, and so forth. Consequently, it is difficult to establish centralized control and determine the actual amount of major expenditures.

The need to rectify this problem led to the creation of program budgeting, where income and expenditures are aggregated across projects or programs. This aggregation is usually in addition to the aggregation done across organizational units. Each project has its own budget, which is divided by task and the expected time of completion.[2]

4.5.1 Time-phased Budgets

To achieve effective cost control, time phasing of project work is absolutely critical. **Time-phased budgets** allocate costs across both project activities and the anticipated time in which the budget is to be expended.[3] In essence, time-phased budgets consolidate the project budget and the project schedule.

A sample time-phased budget is illustrated in Table 4.3. In this illustration, the aggregate budget for a project activity is broken down across different time periods to reflect planned work on its various phases. Consequently, the time-phased budget is an excellent project control mechanism, because it allows the project manager to determine, during various stages of the project life cycle, how much of the budget monies have actually been expended. This figure can then be compared with the budgeted amount.

A time-phased budget also enables the project team to compare the schedule baseline with the budget baseline, which facilitates the identification of milestones for both schedule performance and project expenditures. Finally, as we will discuss in a later chapter, time-phased budgets are an important component of the earned value management technique.

Table 4.3 Sample of a Time-phased Budget

	Months					
Activity	*January*	*February*	*March*	*April*	*May*	*Total by activity*
Design	3,000	2,000				5,000
Blueprint		5,000	3,000			8,000
Engineering		2,000	2,500	1,000		5,500
Test			6,500			6,500
Sequencing			4,000	7,000	1,000	12,000
Production analysis				3,000	2,000	5,000
Monthly planned	3,000	9,000	16,000	11,000	3,000	
Cumulative	3,000	12,000	28,000	39,000	42,000	42,000

4.5.2 Tracking Chart

Once a time-phased budget has been established, we can produce a tracking chart that illustrates expected budget expenditures by plotting the cumulative budgeted cost of the project against the baseline schedule. Figure 4.3 shows a simplified example of the plot, which is another method for identifying the schedule and budget baseline over the anticipated life of the project.

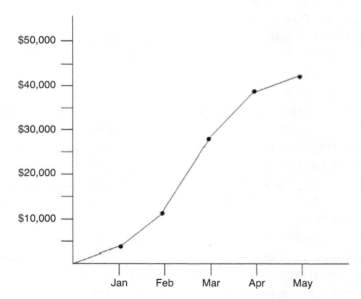

Figure 4.3 Cumulative budgeted cost of the project

4.6 DEVELOPING A PROJECT CONTINGENCY BUDGET

A **project contingency budget** is created to offset project uncertainties, such as unforeseen events that can render initial budgets inaccurate and meaningless. An example of this is a construction project with a fixed amount for digging a building foundation. During excavation, the construction crew encounters previously undetected ground-water problems that must be corrected before proceeding—even though doing so will have serious negative repercussions to the project budget.

Budget contingencies recognize that project cost estimates are just that: estimates. Even in circumstances where unknowns are kept to a minimum, there is simply no such thing as full knowledge of events. As a result, project teams routinely develop a contingency budget that allocates an extra amount to cover the unpredictable. As with cost estimation, the amount earmarked for contingency funds varies with the level of uncertainty—the higher the degree of uncertainty associated with the project, the greater the amount of contingency allotted.

The purpose of including contingency funds in the budget is simply to ensure that unforeseen events do not delay project completion. They can also be used to offset errors associated with estimation, minor design changes, and other omissions. They increase the chance of work being done within the stipulated amount, which in turn increases confidence in project success. If that confidence can be raised to a point where the amount of contingency appears to be both realistic and achievable, its value as a control tool increases significantly.

4.6.1 Allocation of Contingency Funds

Contingency money can be added to individual project activities, work packages, or the project as a whole. Contingency fund allocation is not part of the activity-based costing process, but instead is over and above the calculated project cost.

Some organizations adopt a formal approach to the allocation of contingency funds and include a structured buildup within the project budget, as shown in Figure 4.4. Individual task budgets are set at a level that is reasonable for the defined work, in the view of the estimator and the manager responsible for the task. This becomes the budget issued to the personnel required to perform the task.

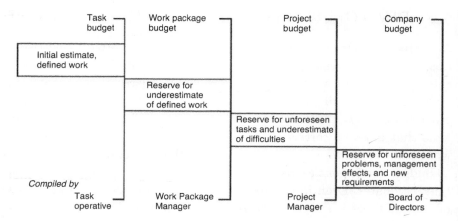

Figure 4.4 Buildup and allocation of contingency funds to various budgets

It should be noted that several minor tasks can come together to form a major task or work package. At this point, by either a rule-of-thumb or a risk-assessment method, a contingency is added to the budget to cover underestimates due to unforeseen difficulties in all the minor tasks. The work package manager may, in theory, control the use of contingency to accommodate any overruns.

4.6.2 Drawbacks of Contingency Funding

Contingency funding is contentious for many project organizations. While project teams support contingency funding as a viable tool for effective project cost control, its acceptance by project stakeholders—particularly by senior management and clients—is a thorny issue.

Clients often feel that contingency funds represent additional expenses for what they interpret as poor control and mismanagement of the original project budget. Furthermore, many clients also feel that the methods used to calculate contingency funds are somewhat arbitrary. For example, it is not uncommon in the building industry to apply a contingency rate of 10 to 15 percent to any structure prior to architectural design. As a result, a building budgeted for $10 million dollars would be designed, at best, to cost $9 million, with the additional million dollars not applied to the operating budget. Instead, it is held in escrow as contingency against unforeseen difficulties during construction.

Another point of contention with contingency often arises over the decision of where (which project activities) contingency should be applied. Does the contingency fund apply equally across all project work packages, or should it be held in reserve to support only a smaller set of critical activities? Clearly, the project team member whose activities receive additional contingency funding has an advantage over other team members receiving no monetary buffer.

A final drawback of contingency budgeting is its vulnerability. All too often, for example, senior management on either the company or customer side can conclude that the contingency estimate is too high. They assume that project planners and managers have made pessimistic assumptions that problems of the past will happen again. Further, in a business environment where it is common to "tweak" budgets to minimize expenditures, contingencies are a natural target for up-front budget revisions. They may be seen as a luxury provision for something that may never happen, and are eliminated to bring the budget down to an "acceptable" level.

The problem of vulnerability can be solved if project stakeholders deal in an authentic manner with the project planners who initially added the contingency. It is critical for top management to work in partnership with the project team to identify reasonable contingency levels, as well as uncertain or risky activities requiring contingency, and then to support the use of appropriate levels of contingency funding.

4.6.3 Advantages of Contingency Funding

In spite of these drawbacks, there are several advantages to the use of contingency funding for projects. First, there is explicit recognition that the future is fraught with uncertainty, and that there is potential for the occurrence of risk events that can have a negative impact on the project budget. Because of this, spending over and above the amount allocated for the project is more common than underutilizing it. Without contingency funds, which act as a cushion against time and money variances, project delays typically mean late completion and a continuous drain on budget.

Second, very few projects are completed within their budgets, and many overrun their targets—often by quite large amounts. This does not necessarily imply poor control, but rather stems from the impossibility of anticipating all that will happen in the future. Through contingency funding, provision is made for these likely, but unexpected, increases in project cost.

Third, the use of contingency funds in a project is an early warning signal of a potentially overdrawn budget. When contingency funds are being applied to the project, there is a clear message that the normal operating budget has been spent, and there is a need for an alternative funding source. In the event of such signals, the organization's top management needs to take a serious look at the project, examine the reasons for its budget variance, and begin formulating alternative plans should contingency funds prove to be insufficient to cover the project overspend.

4.7 ISSUES IN BUDGET DEVELOPMENT

Developing a project budget is a complicated, time-consuming process that requires a number of issues be addressed simultaneously. To do it correctly, we must maintain our focus on three important points:

1. The project team not only has to determine which categories of costs are relevant and appropriate for the project, but also has to identify the principle cost driver (labor, material, etc.) for each project activity, so that a reasonable expense can be charged against that activity. Furthermore, in allocating these resources to the project activity, the project team has to ensure that the task is completed within the time allocated to it by the project schedule.
2. The project team has to decide on the amount of contingency funds to hold in reserve, which is shown as a separate line item in the project budget. The issue of contingency is problematic, as there is usually a fine line between appropriate safety and too much or too little caution. The key lies in finding a reasonable middle ground: one that neither overstates nor underestimates the need to build in some measure of contingency.
3. If the project falls behind schedule, the project team has to make a decision regarding accelerating (crashing) project activities, which can have serious implications for the project budget. In the remainder of this chapter, we will discuss some of these important issues of project acceleration, budgeting, and their ramifications on overall project cost.

4.8 CRASHING THE PROJECT: BUDGET EFFECTS

Project acceleration, or "**crashing**," involves shortening activity duration times by adding resources and incurring additional direct costs, both of which have a direct and significant impact on the project budget. To illustrate this point, Figure 4.5 reproduces a typical time–cost curve that highlights the relationship between accelerating a project and its budget implications. It is clear, from this figure, that there is an inverse relationship between the cost of crashing and the time saved in accelerating an activity's schedule.

Because of this, the decision to crash project activities can never be taken without serious consideration of the consequences. What kind of schedule savings can be realized? How much will it cost? Will top management support any significant budget overruns in return for maintaining the schedule? These are just some of the critical issues that must be resolved prior to any decision to move forward with crashing project activities.

Ultimately, if the cost of crashing an activity is disproportionately high for the time saved, then it is not advisable to crash that activity. On the other hand, if time and schedule issues are critical, as in the case of event project management like the Olympic Games, it is acceptable to incur the extra cost and the decision to crash the project should be supported.

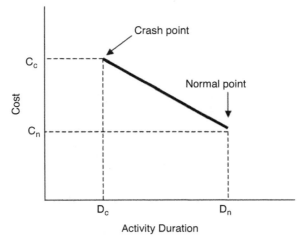

Figure 4.5 Standard time–cost trade-off curve

Crashing Project Activities—Decision Making

Before making a decision to crash project activities, project managers need to make a careful examination of the project budget. This will help them to determine

1. Which activities are the most likely candidates to be crashed
2. The additional costs related to the decision to crash these activities
3. The impact on the overall budget, including a comparison to the time gained from the decision to crash activities

To illustrate the considerations involved in crashing projects, consider the following example. Table 4.4 illustrates a simple project with eight activities. The table identifies the normal duration and associated cost for each of the activities. Further, it estimates how much each of these activities could be shortened, as well as the associated costs of crashing them.

The first decision relates to the candidates for crashing. To shorten project duration, at least one of the activities to be crashed must be on the critical path. Using Table 4.4, let's assume that critical path activities are A, C, D, and H. A simple, side-by-side comparison of these activities and their crash costs is shown in Table 4.5.

Using Table 4.4, we can evaluate all candidates for crashing in terms of time gained versus cost to the project budget. First, we find that in crashing activity C, we save three days at a cost of $1,500 in extra expenses, which makes it the least expensive option. Crashing activity A saves the project three days at an additional cost of $2,000, raising

Table 4.4 Project activities, durations, and direct costs

Activity	Normal		Crashed	
	Cost	Duration	Extra cost	Duration
A	$2,000	10 days	$2,000	7 days
B	$1,500	5 days	$3,000	3 days
C	$3,000	12 days	$1,500	9 days
D	$5,000	20 days	$3,000	15 days
E	$2,500	8 days	$2,500	6 days
F	$3,000	14 days	$2,500	10 days
G	$6,000	12 days	$5,000	10 days
H	$9,000	15 days	$3,000	12 days

Note. Activities on the critical path are highlighted in bold.

Table 4.5 Crash costs for critical activities

Activity	Crash cost
A	$2,000
C	$1,500
D	$3,000
H	$3,000

the total cost of A to $4,000. Crashing activities D and H represents a timesaving of five and three days, respectively, at additional costs of $3,000 each.

Next, every time we determine to crash a project activity, it is necessary to reexamine the project master schedule for its effect on the overall project and other noncritical feeder paths. This is because the decision to crash some activities can sometimes have the effect of raising other activities to the critical path. In the above example, if activity C (of 12 days' duration) were crashed, the project schedule would need to be reexamined to determine if saving three days on this activity would lengthen a noncritical path.

The next decision to be made is whether to crash multiple times, because each attempt often leads to different choices for the best activity to be crashed. For example, if the results of the original estimate suggested a project schedule of 57 days (the combination of the durations of activities A, C, D, and H), the first crash decision shrank the schedule to a 54-day baseline at a cost of $1,500. It is possible to repeat this process—identifying the lengths of critical-path activities, weighing cost trade-off decisions, and continuing to shrink the durations of project activities—until there are no more to crash. Those activities that do remain do not affect the critical path, and, therefore, cannot impact the project's final expected duration.

Through this process, the project team gains time, but also incurs extra direct costs. It is also necessary to account for indirect costs, such as overhead expenses and liquidated damages. The following section illustrates the choices the project team is faced with as they continually adjust the cost of crashing the schedule against other project costs.

In terms of overhead, suppose a project is being charged a fixed rate of $200 per day. Let us further assume that a series of penalties, including liquidated damages, is due to kick in if the project is not completed within

Table 4.6 Project costs over duration

Project duration (in days)	Direct Costs	Liquidated damages penalty	Overhead costs	Total costs
57	$32,000	$5,000	$11,400	$48,400
54	33,500	3,000	10,800	47,300
51	35,500	1,000	10,200	46,700
48	38,500	0	9,600	48,100

50 days. The original 57-day schedule clearly created a risk for penalties. Further, the first crashed schedule, while improving the delivery date, was still four days over the deadline. However, suppose we discover that iterating the crashed schedule three times will take us from our original 57-day schedule to a new schedule of 48 days (crashing first activity C, then A, then H). The schedule has shortened nine days against a budget increase of $6,500 (see Table 4.6).

We complete this table by following the costs for each successive crashed activity and linking them to total project costs. Intuitively, we can see that direct costs will continue to increase as we include the extra costs associated with additional iterations of crashed activities. On the other hand, overhead charges and liquidated damages costs would decrease; in fact, at the 48-day mark, liquidated damages no longer factor into the cost structure. Here, the decision becomes simply one of at what point it is no longer economically viable to continue crashing project activities.

Figure 4.6 offers a visual explanation of the choices the project team made in balancing the competing demands of schedule and cost, particularly when other intervening factors were included, such as penalties for late delivery. Direct costs are shown with a downward slope, reflecting the fact that costs will rapidly ramp up as the schedule shrinks (the time–cost trade-off effect). However, if we also allow liquidated damage penalties to emerge after the 50-day schedule deadline, we see that the project team is facing a choice of paying extra money for a crashed schedule at the front end, versus paying out extra penalties for delivering the project delivery past the scheduled delivery date. The goal here is to minimize total project cost, represented as a balancing act between competing cost drivers—crashing costs and the cost of liquidated damages.

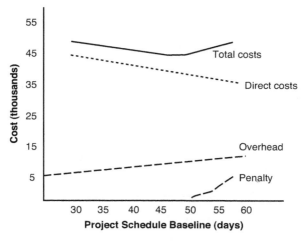

Figure 4.6 Budgetary costs of crashing a project
Source: Shtub et al.[4]

In summary, the important points regarding decision-making in project crashing are as follows:

- To shorten project duration, at least one of the activities to be crashed must be on the critical path.
- The activity to be crashed first is the one that has lowest marginal cost of crashing, compared to the other activities on the critical path.
- Crashing can occur multiple times, and the process can eventually lead to multiple critical paths. In such cases, the activities to be crashed are chosen from each of the critical paths that have the lowest marginal cost of crashing.
- In addition to the direct costs, it is also necessary when crashing a project to account for indirect costs such as overhead expenses and liquidated damages.

Along with project cost estimation, project budgeting provides the basis for establishing sound project control and profitability. To create an accurate budget for a project, we need to understand the difference between top-down and bottom-up budgeting, including their advantages and disadvantages. Further, because budgeting is truly only as good as the approaches that were used to first estimate costs, a better understanding of some of the key processes for cost estimation are required. Finally, the budget baseline must work in relation to the project schedule. This necessitates the creation of a time-phased budget that recognizes the sequencing of project activities and allows the project team to identify their budget, including assessing its status on an ongoing basis. When properly managed, the budget, working together with the schedule, offers the project team the opportunity to apply maximum control over the project.

REFERENCES

1. Maher, M., Lanen, S., Anderson, S., and Maker, M. (2010) *Fundamentals of Cost Accounting*, 3rd ed., Chicago, IL: Irwin.
2. Meredith, J. R., and Mantel, S. J., Jr. (2011) *Project Management—A Managerial Approach*, 8th ed., New York: Wiley.
3. Project Management Institute, Inc. (2011) *Project Management Body of Knowledge*, 4th ed. Newton Square, PA: PMI.
4. Shtub, A., Bard, J. F., and Globerson, S. (2004) *Project Management: Engineering, Technology, and Implementation*, 2nd ed., Englewood Cliffs, NJ: Prentice–Hall.

KEY TERMS

Project budgets
Fiscal operating budgets
Top-down budgeting
Bottom-up budgeting

Activity-based costing (ABC)
Time-phased budgets
Project contingency budget
Project acceleration—"crashing"

Chapter 5

Project Cost Control

LEARNING OBJECTIVES

- Describe project control process.
- Demonstrate how to conduct an earned value analysis.
- Demonstrate how to perform an earned value assessment to individual projects or a portfolio of projects.
- Discuss the important issues in the effective use of EVM.

In Chapters 3 and 4, we examined the first two essential aspects of cost management: cost estimation and project budgeting. The next important part of this process is **project cost control**, which typically falls under the larger umbrella of project evaluation and control. For this reason, any discussion of project cost control must begin with a brief overview of the larger framework. This is followed by an examination of the various aspects and approaches to evaluating and controlling project costs, including S-curve analysis of time–cost relationships. The chapter concludes with an in-depth look at another valuable mechanism for controlling project costs called earned value management (EVM).

5.1 OVERVIEW OF THE PROJECT EVALUATION AND CONTROL SYSTEM

A project evaluation and control system measures project progress and performance against a project plan to ensure that the project is completed on time, within budget, and to the satisfaction of the customer. A good project evaluation and control system should also provide project managers with advance warning of potential problems before it is too late to correct them. Without these systems, projects proceed aimlessly with very little oversight, without a clear understanding of status, and without a well-thought-out action plan to bring the project back on track in the event of obvious disruptions.

Designing, implementing, and maintaining an accurate monitoring and control system is perhaps one of the most difficult challenges in project management, and more than a few organizations get it wrong. There are two reasons for this. First, very few project managers and project teams have a strong grasp on the essentials of project control. Subsequently, they do not know what warning signals to look for during the development process, or when to look for them. Second, they do not have the necessary know-how and training to develop systematic project control that is comprehensive, precise, and timely. The good news is that this can be easily remedied with some basic knowledge of the processes and procedures involved.

5.1.1 Project Control Process

To correctly and accurately measure and evaluate project performance, four essential elements must be in place. They include setting a baseline plan, measuring progress and performance, comparing actual performance against plan, and taking corrective action—each of which is now examined in detail.[1]

1. *Establishing a project baseline plan*—The project baseline plan provides the essential features for measuring performance. It begins with an accurate **work breakdown structure (WBS)**, which establishes all the work packages and tasks associated with the project, assigns the personnel responsible for them, and creates a hierarchical representation of the project from the highest level down. To create the project baseline plan, the project team lays out each of the discrete tasks from the WBS onto a project network diagram, and time-phases all work, resources, and budgets.
2. *Measuring and monitoring progress and performance*—Accurate mechanisms for project measurement are essential prerequisites of effective control systems. The first step in creating them is to establish a control system that measures the ongoing status of various project activities in real time, and provides project managers with relevant information as quickly as possible.

    The second step is to determine what should be measured. There are both quantitative and qualitative measures for monitoring project progress, and integrating quantitative measures like time and cost into the control system is relatively easy. On the other hand, qualitative measures like customer satisfaction with product functionality and technical specification can be determined only through on-site inspection or actual use.

When it comes to quantitative measurement, evaluating project performance relative to time can be as simple as answering questions like, "Is the critical path early, late, or still on schedule?" or, "Is there reduction in the slack of noncritical paths?" Measuring project performance against budget (e.g., dollars, units in place, labor hours) is more difficult. In these cases, a form of project measurement known as **earned value management (EVM)** can provide a realistic estimate of project performance against a time-phased budget. We will examine EVM later in this chapter.

3. *Comparing actual performance against plan*—Given that actual project performance is rarely in accordance with the original baseline plan, the next step is to compare the two to measure deviations. This analysis—sometimes referred to as **"gap" analysis**—is essential for determining current project status. As a rule, the smaller the deviation between the baseline plan and actual performance, the easier it is to take corrective action.

4. *Taking remedial action*—In cases where the deviations between the plan and actual performance are large and obvious, some form of corrective action is necessary to bring the project back on track. In some cases, the action may be relatively minor; in others, it may require serious and significant remedial steps. In situations where conditions or project scope have changed, the original baseline plan may have to be revised.

Finally, it is important to note that this monitoring and control process is not a one-time fix, but is a continuous cycle of goal setting, measuring, correcting, improving, and remeasuring as illustrated in Figure 5.1.

5.2 INTEGRATING COST AND TIME IN MONITORING PROJECT PERFORMANCE: THE S-CURVE

Gantt charts, control charts, and milestones are tools that are often used to monitor project performance. However, these tools track progress only in the dimension of time. The other important dimension of project performance, cost, is virtually ignored.

PROJECT COST CONTROL

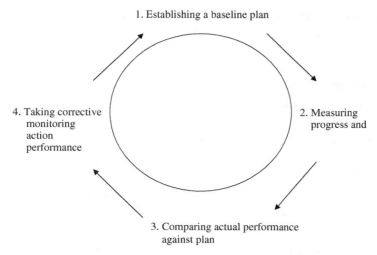

Figure 5.1 The project control process[2]

One of the mechanisms that monitors both dimensions is time–cost analysis. In this method, the project's progress is monitored as a function of the cumulative costs and plotted against time for both budgeted and actual amounts. **Time–cost analysis** can be illustrated through the simple example of a fictional project, "Project Orion," which consists of four work packages (design, engineering, installation, and testing). The completion budget is $100,000, and the anticipated duration is 50 weeks. A breakdown of the project's cumulative budget, in terms of both work package and time, appears in Table 5.1.[3]

In time–cost analysis, the relationship between time and money is represented graphically with time on the x, or horizontal axis, and money spent on the y, or vertical axis. The typical form of this relationship is S-shaped, where budget expenditures are initially low and increase rapidly during the

Table 5.1 Budgeted costs (in $ 000) for Project Orion

	Duration in Weeks										
Work package	5	10	15	20	25	30	35	40	45	50	Total
Prototype design	6	2	4								12
Engineering		4	8	8	8	6					34
Assembly				4	20	6	6				36
Test						2	6	4	2	4	18
Total	6	6	12	12	28	14	12	4	2	4	100
Cumulative	6	12	24	36	64	78	90	94	96	100	100

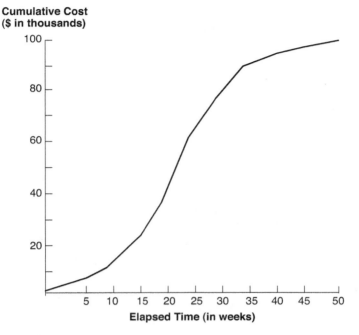

Figure 5.2   Project S-curve

major project execution stage before starting to level off again as the project gets nearer to completion. In Figure 5.2, the cumulative budget projections for Project Orion shown in Table 5.1 have been plotted against the project's time duration. The S-curve figure represents the project budget baseline against which actual budget expenditures will be evaluated.

To monitor the status of a project using an S-curve, the cumulative project budget expenditures to date are compared with the actual spending patterns at the end of each time period of interest. Any significant deviations between actual and planned budget expenditures constitute a potential problem area that must be investigated.

S-curves were an early attempt to illustrate the nature of the important relationship between time and money. They help project managers understand the correlation between project duration and budget expenditures and give them a good sense of where the highest levels of budget spending are likely to occur. Using S-curves to monitor the status of a project involves the following: At the end of each time period of interest, the cumulative project budget expenditures to date are compared with the actual spending patterns. Any significant deviations between actual and planned budget expenditures would constitute a potential problem area that must be investigated.

The biggest advantage of S-curve analysis is that it is simple to use and data can be created and presented without much difficulty. Furthermore, the S-curve provides the most current information on the project status as budget expenditures can be constantly revised and the new value plotted on the graph. The S-curve is a more proactive control mechanism, as information can be represented immediately and updated continuously. Consequently, the S-curves provide a clear visual picture of the project's status that is available in a timely manner and easy to read.

To illustrate S-curve analysis using the example in Table 5.1, we'll assume that as of week 27, the original projected expenditure is $75,000. However, the actual project expenditures totaled only $65,000. The net effect is that there is a $10,000 budget shortfall, or negative variance between the cumulative budgeted cost of the project and its cumulative actual cost. Figure 5.3 shows this comparison of budget versus actual costs, along with the negative variance as of week 27.

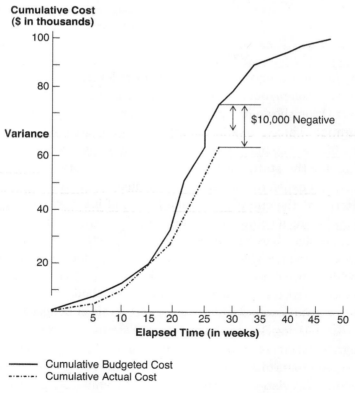

Figure 5.3 Project S-curve showing negative variance

While S-curves provide project teams with many benefits, the greatest is that S-curve analysis is simple to use, and data can be created and presented without much difficulty. In addition, S-curves

- Help project managers understand the correlation between project duration and budget expenditures, and provide a good sense of where the highest levels of spending are likely to occur.
- Provide the most current information on project status, because budget expenditures can be constantly revised and new values plotted on the graph.
- Serve as a more proactive control mechanism, because information can be immediately represented and continuously updated.

Despite these advantages, S-curves have a number of significant shortcomings that must be taken into account when project teams contemplate their use. For example, while S-curves can identify deviations between actual and budgeted expenditures (both positive and negative variances), the cause of these deviations cannot easily be determined. In the S-curve shown in Figure 5.3, it is clear that the budgeted amount as of week 27 has not been expended. However, what we do not know and cannot conclude from this graph is the reason for the negative variance. Is it an indication that the project is behind schedule, or that the project team has come up with more efficient methods of completing the tasks? In either case, there is potential for misusing S-curves as a project-monitoring tool.

In the final analysis, simply evaluating a project's status vis-à-vis its performance on time versus budget expenditures can easily lead to erroneous conclusions about project performance.[4] This disadvantage created significant problems for several high-profile U.S. aircraft development projects in the early 1960s, and ultimately led to the adoption of the more popular analytical approach of earned value management.[5]

5.3 EARNED VALUE MANAGEMENT

The remainder of this chapter focuses on **earned value management**, a mechanism that can determine how much work was accomplished for the money spent. The earned value system uses the data from the work breakdown structure, the project network, and the schedule to compare time-phased costs with scheduled activities. In the process, it enables meaningful comparisons to be made between actual and planned schedules and costs.

The use of EVM as a project monitoring and control mechanism began in the 1960s, when it was championed by the U.S. Government as a viable system for its agencies and contractors to track project performance. The focus was to track the "value" performance of projects, in addition to cost and other traditional measures. In the 30 years since its origin, EVM has been used worldwide in a wide variety of settings, ranging from governmental agencies to a host of project-based organizations in numerous industries.

Unlike Gantt charts and S-curves, EVM evaluates a project by integrating the criteria of time, cost, and *value*. In other words, in addition to comparing actual and budgeted costs, EVM integrates the important element of time in determining what was accomplished (value realized) to draw conclusions about current project status. In essence, the earned value method measures the *value of the work actually accomplished at the cost rates set out in the original budget*. This quantity is known as **earned value (EV)**. Furthermore, as EVM provides information about the efficiency with which budgeted money is used relative to the value realized, forecasts about the estimated cost and schedule to project completion can be made.

5.4 EARNED VALUE MANAGEMENT MODEL

Earned value management shows the relationship among all three of the primary project success criteria: cost, schedule, and performance. The figures below illustrate the superiority of EVM in comparison with the

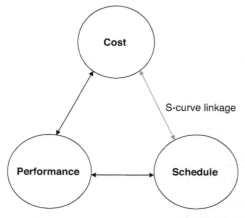

Figure 5.4   Project performance dimensions linkage in S-curve analysis[6]

Earned Value Management Model 113

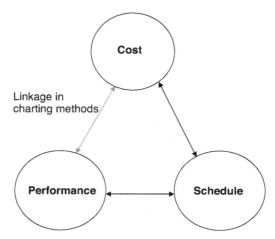

Figure 5.5 Project performance dimensions linkage in charting methods[7]

other project-tracking mechanisms, such as Gantt charts and S-curves. Essentially, S-curves establish a linkage directly and solely between cost and schedule (Figure 5.4). Tracking mechanisms, such as tracking Gantt charts, employ links between schedule and project (or activity) performance (Figure 5.5). It is only through earned value that the full nature of the association between the three success metrics of schedule, cost, and performance can be understood in relation to each other (Figure 5.6).

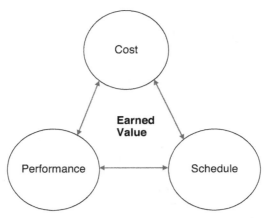

Figure 5.6 Project performance dimensions linkage in EVA[8]

5.5 FUNDAMENTALS OF EARNED VALUE

The two key elements involved in developing an earned value analysis are

- *The work breakdown structure*—This provides, in a hierarchical structure, information regarding the individual tasks that need to be performed on the project and individual work packages. The WBS makes it possible to allocate necessary human resources that match task requirements. Subsequently, the project network derived from this information enables the correct sequencing of tasks to be identified, and provides the basis for developing a time-phased budget.
- *A time-phased budget for each work package*—With a time-phased budget in place, the project team can determine the timing of budget expenditures required to complete individual tasks. More importantly, the time-phased budget enables the project team to determine the points in the project when budget money is likely to be spent in pursuit of those tasks.

As an example, let's assume that the design activity for a project has a budgeted amount of $100,000 and requires four months to complete. Let's further assume that the major portion of the design work will be completed in the first three months. A time-phased budget for this activity may look like Table 5.2.

Now that we have the task and work package information from the WBS and applied a time-phased budget breakdown, the project baseline can be developed.

5.6 EVM TERMINOLOGY

The standard EVM terminology currently in use (with some minor variations) was first devised by the U.S. Department of Defense, and is as follows[9]:

- *PV, planned value*—Comprises of all the relevant costs in the project, or any given part of the project, up to the reporting date.

Table 5.2 Time-phased budget for design activity

Activity	Jan	Feb	Mar	Apr	May–Dec	Total
Design	$20,000	$40,000	$35,000	$5000	0	$100,000

- *EV, earned value*—The cost of all progress achieved on the project, or part of the project, up to the reporting date, expressed in terms of the costs originally set out in the initial estimate. It represents what has been earned, not simply what has been spent.
- *AC, actual cost of work performed*—The cumulative expenditures on the project, or part of the project, up to the reporting date.
- *CV, cost variance*—Given by (EV − AC).
- *SV, schedule variance*—Given by (EV − PV).
- *CPI, cost performance index*—Given by (EV/AC).
- *SPI, schedule performance index*—Given by (EV/PV).
- *OD, original duration.*
- *ETC, expected time to completion*—Given by (OD/SPI)
- *BAC, budgeted cost at completion*—Represents the total budgeted cost of the project baseline.
- *EAC, estimated cost at completion*—Represents the sum of the costs incurred to date and the revised estimated costs for the work remaining, given by (BAC/CPI).
- *FAC, computed forecasted costs at completion.*
- *VAC, variance at completion*—Given by (BAC − EAC) or (BAC − FAC), indicates expected positive or negative deviation at completion.

5.7 RELEVANCY OF EARNED VALUE MANAGEMENT

To illustrate the relevancy of earned value, we'll return to the Project Orion example. Recall that this project consists of four work packages (prototype design, engineering, assembly, and testing), a budget to completion of $100,000, and an anticipated duration of 50 weeks. (For more information, please refer to Table 5.1 and the associated S-curve in Figure 5.2.)

Assuming that the projected project costs and actual expenditures are the same, we can determine that the project budget is being expended within the stipulated timeframe as of week 30. However, with some revised information, Table 5.3 shows the actual status of the hypothetical Project Orion. An examination of this table reveals that as of week 30, work packages related to prototype design and engineering have been fully completed, assembly is 50 percent done, and testing hasn't begun. (Percentage values like those in Table 5.3 can be obtained from the

Table 5.3 Project Orion's budgeted costs (in $000) and percentage completed

Work package	Duration in Weeks										Total	% Completed
	5	10	15	20	25	30	35	40	45	50		
Prototype design	6	2	4								12	100
Engineering		4	8	8	8	6					34	100
Assembly				4	20	6	6				36	50
Testing						2	6	4	2	4	18	
Total	6	6	12	12	28	14	12	4	2	4	100	
Cumulative	6	12	24	36	64	78	90	94	96	100	100	

project team or a knowledgeable individual's assessment of the current status of work package completion.)

At this point, the earned value of the project work completed thus far is the critical question that needs to be answered. In other words, from the perspectives of budget, schedule, *and* performance, what is the status Project Orion as of week 30?

Based on the information in Table 5.2, the earned value for these work packages can be easily calculated by totaling the product of the planned budget for each work package and its respective percentage completed. The resulting sum is the earned value to date for both the work packages and the overall project. For the Project Orion example, earned value (as shown in Table 5.4) at the end of 30 weeks is $64,000.

The earned value calculated above can now be compared with the planned budget, using the original project budget baseline. The results are shown in Figure 5.7, which provides a more realistic assessment of project status than the S-curve shown in Figure 5.3. According to the original budget projections, $68,000 should have been spent as of week 30. The earned value analysis, however, is projecting a shortfall of $17,000. In other words, there is a negative variance not only in

Table 5.4 Earned value calculations (values in thousands $)

Work package	Planned budget	Percentage completed	Earned value
Prototype design	12	100	12 * 1 = 12
Engineering	34	100	34 * 1 = 34
Assembly	36	50	35 * 0.5 = 18
Testing	18	0	18 * 0 = 0
Cumulative earned value			64

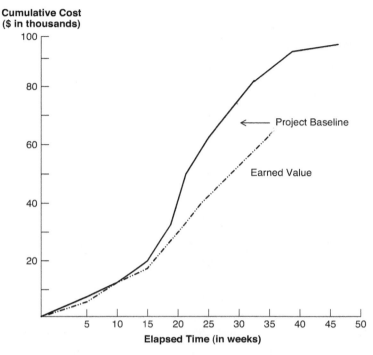

Figure 5.7   Project baseline comparison using earned value

terms of money spent on the project, but also in terms of the extent of value created (performance) in the project to date. Specifically, while the $10,000 negative variance shown in Figure 5.3 may or may not be of serious consequence, a $17,000 shortfall in value earned on the project to date is certainly a cause for concern.[10]

5.8   CONDUCTING AN EARNED VALUE ANALYSIS

There are several important steps to be considered when conducting an earned value management analysis, including the following[11]:

1. *Clearly define each project activity or task, its resource needs, and a detailed budget for that task*—Using the work breakdown structure, the project team can identify all necessary project tasks and the project resources assigned to that task, including personnel assignments and costs for equipment and materials. Given these task breakdowns and resource assignments, the budget or cost estimate for each project task can be generated.

2. *Develop schedules for activity and resource usage*—The purpose of this step is to determine, on a period-by-period basis, the percentage of the total budget allocated to each task throughout the life of the project. The outcome of this process is the establishment of a direct link between the project budget and the project schedule. Specifically, this step provides information on how much of the budget is allocated to each task, along with when the resources will be used during the project development cycle.
3. *Develop a "time-phased" budget*—Armed with the information available from the previous step, it is now possible to establish expenditures across the project's life. With this information, we can now determine the planned value, which is the total (cumulative) amount of the budget, and serves as the project baseline. During any stage of the project development cycle, the PV helps to identify the cumulative budget expenditures planned for that stage. The PV in any period is a cumulative value, which is sum of all planned budget expenditures of all preceding time periods.
4. *Determine and aggregate the actual costs incurred for each task that is being performed*—The aggregate actual costs of performing a task represent the actual cost of work performed (AC). In addition, we can also calculate the budgeted values for the tasks being performed. This gives us our project's earned value (EV).
5. *Compute the cost and schedule variances as the project progresses*—The three key pieces of data collected from the previous steps (PV, AC, and EV) can be used to calculate both the project's cost variance and schedule variance while the project is still in progress. The schedule variance is calculated as the difference between the budgeted cost of the work performed and the budgeted cost of the work scheduled to be performed to date, and is given by $SV = EV - PV$. The budget, or cost, variance is defined as the difference between the budgeted cost of the work performed and the actual cost of the work performed to date, and is given by $CV = EV - AC$.

Figure 5.8 presents a simple model that integrates the three key components of earned value (PV, AC, and EV). The project baseline (PV), which represents both the schedule and budget for all project tasks, is shown in the bottom left corner of the chart. Any actual schedule deviation from the original PV is attributed to the EV. Given these earned value figures, which indicate the extent to which project tasks are completed,

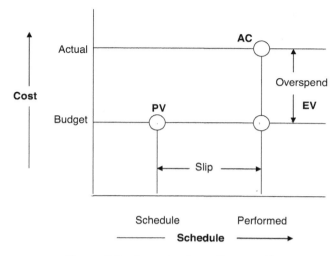

Figure 5.8 Earned value milestones[12]

we can now compute the project's AC, which is the difference between the budgeted and actual costs of the project's activities.

5.9 PERFORMING AN EARNED VALUE ASSESSMENT

Table 5.5 presents the initial components of a calculated EVA on a fictitious project, titled "Project Jupiter."[13] This project, which began in January, has a planned duration of eight months and a $150,000 budget.

We are interested in calculating the earned value of Project Jupiter at the end of July. For purposes of simplicity, we will assume that the project comprises only six work packages. If we know the amount budgeted for each work package and when the work is scheduled for completion, we can begin to construct a budget table like the one shown in Table 5.5. This table shows that each work package has a fixed budget that runs across several time periods. For example, "Preliminary design" is budgeted to cost $20,000, and will be performed during the months of March and April. "Prototype development" is scheduled to begin in April, has a budgeted amount of $5,000, and concludes in May.

In Table 5.5, we can see the amount budgeted for each work package for each completed month of the project (January through July). In addition, we can see the actual amount spent each month in the bottom four rows. Although we had planned to spend $40,000 on activities by April, the actual cumulative costs were $45,000. On the surface, it appears that

Table 5.5 Earned value table for Project Jupiter

Activity	Jan	Feb	Mar	Apr	May	Jun	Jul	Aug	Plan	% complete	Value
Staffing	9	6							15	100	15
Preliminary design			8	12					20	80	16
Prototype development				5	10				15	60	9
Testing					5	5			10	60	6
Final design					2	8	15		25	40	10
Construction						5	40		45	20	9
Transfer								20	20	0	0
Total									150		65
Monthly plan	9	6	8	17	17	18	55	20			
Cumulative	9	15	23	40	57	75	130	150			
Monthly actual	9	11	10	15	15	18	40	0			
Cumulative actual	9	20	30	45	60	78	118				

the budget has been overspent—but, as noted earlier, this information is insufficient to arrive at any realistic determination of project status.

What has to be determined is the value earned on the project to date; that is, the number of work packages and the percentage of work completed in each, over the time budgeted to them. The key pieces of information that allow us to identify earned value are included in the three columns at the far right of Table 5.5. These columns show the planned expenditures for each work package, the percentage of work completed in each, and the calculated "value," which is simply the product of the planned expenditures and the percentage of work completed.

For example, the work package "Preliminary design" has a total planned budget of $20,000 across two months. To date, however, only 80 percent of that work package has been completed, resulting in $16,000 in "value." In the same manner, if we sum the numbers in the columns for planned expenditures and actual value (EV), we come up with our project's planned budget ($150,000) and the value realized at the end of July ($65,000).

With the information that we now have in Table 5.5, it is possible to use earned value analysis to determine the project's status. In this table, the planned value (PV) is the cumulative cost at the end of the month of July ($130,000). The earned value (EV) for the project to date, totaling $65,000, is at the bottom of the last right-hand column. The performance measures that are of interest to us are the schedule performance

Table 5.6 Schedule variances for EVM

Planned value (PV)		130
Earned value (EV)		65
Schedule performance index (SPI):	EV / PV = 65/130 =	0.50
Estimated time to completion (ETC)	OD / SPI (8/0.50) =	16 months

index (SPI) and the estimated time to completion. The SPI is calculated by dividing the EV by the PV, as shown in the calculation appearing in Table 5.6 ($65,000/130,000 = 0.50).

With the SPI, we can now determine the amount of time required to complete the project. Since the SPI is telling us that we are operating at only 50 percent efficiency, we divide the original project duration (OD) by the SPI to determine the projected actual timeframe for completion (8/0.50 = 16 months). Clearly, as of July, it appears that the project will take an additional nine months; in other words, we are running exactly eight months behind schedule.

Now that we have determined that our project is eight months behind schedule, the next step is to use the EVM to make similar projections about the final cost. The process for computing cost variances is similar to calculating schedule variances, except that the two important pieces of information we need are the actual cost of work performed (AC) and earned value (EV). We already know the earned value figure ($65,000), and we can determine the AC from Table 5.5. This is the cumulative actual cost incurred at the end of July ($118,000).

The cost performance index (CPI) for this project is obtained by dividing the EV by the AC ($65,000/118,000 = 0.55). The projected cost is calculated by dividing the original project budget ($150,000) by CPI (0.55) to arrive at the estimated cost to completion of $272,727 ($150,000/0.55). These calculations are presented in Table 5.7. The disconcerting news is that this project not only is well behind schedule, but is also projected to incur a significant cost overrun.

Figure 5.9 is a graphical representation of these variance values that shows the differences between earned value (EV) and PV and AC.

Table 5.7 Cost variances for EVA

Actual cost of work performed (AC)		118
Earned value (EV)		65
Cost efficiency index	EV/AC = 65/118 =	0.55
Estimated cost to completion	BAC/CPI = ($150,000/0.55) =	$272,727

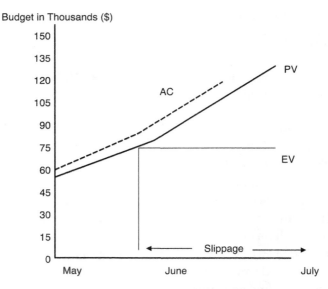

Figure 5.9   Earned value variances

The intriguing result of this example suggests how misleading simple S-curves can be.

For example, in Table 5.5, the difference of $12,000 between the AC ($118,000) and PV ($130,000) at the end of July seems to indicate that we have marginally underspent our budget. However, when viewed from the perspective of earned value ($65,000), the results were much more serious. This is due to the fact that when we calculated the percentage of completion for all scheduled tasks, the shortfall in earned value revealed that the schedule and cost variances were much more severe.

This example clearly underscores earned value management's superiority for precisely determining actual status—which, as we've shown, is a function of its three component pieces: schedule, budget, and extent of project completion.

5.10   MANAGING A PORTFOLIO OF PROJECTS WITH EARNED VALUE MANAGEMENT

Earned value management methodology can be applied to both individual projects and portfolios of projects. With the latter, all of the earned value measures across the firm's entire portfolio of projects are aggregated, with the aggregate earned value figure serving as a measure of the efficiency

Table 5.8 Project portfolio earned value[15] (all figures in thousands ($))

Project	PV	EV	Time var ($)	Var (+)	AC	Cost var ($)	Var (+)	Plan	Est. at completion
Alpha	91	73	18	18	83	10	10	254	289
Beta	130	135	−5	0	125	−10	0	302	302
Gamma	65	65	5	5	75	15	15	127	159
Delta	25	23	2	2	27	4	4	48	58
Epsilon	84	82	2	2	81	−1	0	180	180
Total	395	373			391				986
Total schedule variance: 27					Total cost variance: 29				
Relative schedule variance: 27/395 = 6.84%					Relative cost variance: 29/395 = 7.34%				

with which the company is managing its projects. Table 5.8 is an example of a portfolio-level EVM control table.[14] It presents both positive and negative cost and schedule variances, and, based on these measures, the projected cost to completion of each current project.

Table 5.8 includes not only information on the total positive variances for both the budget and schedule, but also the relative schedule and cost variances expressed as a percentage of the total project portfolio. For example, the company's portfolio of projects is showing average cost and schedule variances of 6.84 percent and 7.34 percent, respectively.

It is evident from the above analysis that EVM provides a company's top management with an excellent tracking and control tool for monitoring a portfolio of projects. Not only does it provide a measure of a company's ability to efficiently run projects, but it also provides the mechanism to compare all projects currently in development, and facilitates identification of both the positive and negative variances that are incurred.[16]

5.11 IMPORTANT ISSUES IN THE EFFECTIVE USE OF EARNED VALUE MANAGEMENT

Earned value management's effectiveness as a performance metric depends on some <u>important factors</u>. The first, and most important, <u>is the availability of highly accurate, up-to-date information on the percentage of work packages completed</u>, which is vital for determining the earned value at any point in time. The accuracy of the calculated earned value hinges on an honest reporting system, as well as the integrity of the project team members and managers.

In prac[tice, there are several d]ecision rules for assigning completion pe[rcentages. The] common methods for determining completion values are

1. *0/100 rule*—This rule assigns a value of zero to a project activity until it is completed. Once the activity is fully complete, then a 100 percent completion percentage is assigned. This rule is best suited for work packages that have very short durations, because it provides virtually no information on the status of the work package on an ongoing basis.
2. *50/50 rule*—This decision rule assigns a 50 percent completion value for any activity that has been started and carries this value until it is completed, at which time the completion value switches to 100 percent. Like the 0/100 rule, this decision rule is most often used for work packages of very short duration.
3. *Percentage complete rule*—With this decision rule, the project manager and team members mutually decide upon a set of completion milestones. For example, these predetermined completion milestones may be 25, 50, 75, and 100 percent, or 33, 67, and 100 percent, or any other set of values. Next, the status of each in-process work package in the project is reviewed and updated. Depending on the extent of completion, the work package may or may not be assigned a new completion value, and, on the basis of this new information, the project's EVM is updated. Clearly, the viability and integrity of this process rests on an honest appraisal of the status of ongoing activities in terms of the actual percentage of the activities completed, regardless of the elapsed time or budget spent.

The percentage complete rule can be problematic, primarily due to controversy surrounding the level of detail used in calculating an activity's completion value. The various gradients of completion must be acknowledged and used by all parties; otherwise, using EVM involves the risk of creating misleading information. In addition, using the percentage completion rule with excessive levels of detail for EVM is essentially meaningless.

For example, let's assume that the earned value analysis of a project uses percentage completion values based on five percent increments (e.g., 5, 10, 15, etc.). From a practical point of view, it is virtually impossible to precisely delineate between five percentage points for most project activities. Consequently, the use of too much detail has the potential to confuse the true status of a project, rather than to clarify it.

The only exceptions to this are for projects where there is prior knowledge of the nature of the task, or where it is possible to accurately measure the amount work completed. In the case of software development projects, for example, where it is possible to judge quite accurately the number of lines of code needed to complete a task, a higher level of detail for task completion percentages can be employed. In these cases, it is also possible to estimate the cost that would be incurred in task completion.

A second issue when establishing accurate or meaningful EVM results is the "human factor" that comes into play when projecting project activity completion. In the interest of looking good for the boss, or due to implicit or even explicit pressure from project managers themselves, the tendency to downplay serious cost problems can arise.

Despite these limitations, EVM is useful for enabling project managers and their teams to gain a better understanding of the "true" nature of project status midstream—specifically from the aspect of cost control.[17] The real-time information provided by EVM can be invaluable in gathering the most up-to-date cost information and in developing realistic and meaningful plans for addressing and rectifying any systematic problems associated with project cost management. Ultimately, these cost management benefits stem from disciplined planning and the availability of metrics that show real variances from plan to generate necessary corrective actions.

In the final analysis, project cost control is fraught with many uncertainties. Therefore, it is imperative that top management take the time to periodically review budget and financial information. By using a formal review process, it is possible to prevent projects from going adrift or the escalation of costs without sanctions. A formal financial review process can also mitigate the risk of cost overruns, and ensure that the project stays on course.

At the completion of the project, it is the project manager's responsibility to ensure that all costs are accrued and accounted for, that a financial balance sheet is produced for audit and signature, and that the financial procedures of the company are adhered to.[18]

REFERENCES

1. Pinto, J. K. (2009) *Project Management: Achieving Competitive Advantage*. Upper Saddle River, NJ: Pearson Prentice Hall.
2. Pinto, J. K. (2009) *Ibid*, p. 411.
3. Pinto, J. K. (2009) *Ibid*.
4. Pinto, J. K. (2009) *Ibid*.
5. UMIST Module 4 workbook.
6. Pinto, J. K. (2009) *Ibid*, p. 418.
7. Pinto, J. K. (2009) *Ibid*, p. 418.
8. Pinto, J. K. (2009) *Ibid*, p. 419.
9. UMIST Module 4 work book, p. 4.5.6.
10. Pinto, J. K. (2009) *Ibid*.
11. Pinto, J. K. (2009) *Ibid*.
12. Pinto, J. K. (2009) *Ibid*, p. 422.
13. Brandon, D. M., Jr., (1998) Implementing earned value easily and effectively. *Project Management Journal*, 29 (2): 11–18.
14. Brandon, D.M., Jr., (1998) *Ibid*.
15. Pinto, J. K. (2009) *Ibid*, p. 427.
16. Pinto, J. K. (2009) *Ibid*.
17. Christensen, D. S. (1998) The costs and benefits of the earned value management process. *Acquisition Review Quarterly*, 5: 373–386.
18. McManus, J. (2006) British Computer Society—Keeping projects under cost control, http://www.bcs.org/server.php?show = conWebDoc.5912

KEY TERMS

Project cost control
Work breakdown structure (WBS)
Earned value management (EVM)
Gap analysis
Time-cost analysis
S-curve
Actual cost of work performed (AC)
Cost variance (CV)

Schedule variance (SV)
Cost performance index (CPI)
Schedule performance index (SPI)
Expected time to completion (ETC)
Computed forecasted cost at completion (FAC)
Variance at completion (VAC)

Chapter 6

Cash Flow Management

LEARNING OBJECTIVES

- Demonstrate how time value of money impacts a project's profitability.
- Demonstrate how to apply discounting to cash flows.
- Describe the various kinds of financial risk and how they can be controlled.

Cash flow is critical to any business, but from a project organization's point of view, effective cash flow management can mean the difference between project success and failure. This is because the amount of cash inflow and outflow, and the timing of these flows during the course of a project, can have a significant influence on project costs and schedule.

In this chapter, we focus on why managing cash flow is vital to project success. You will learn about the various aspects of cash flow management, including payment arrangements and plans that influence cash flow, and will gain an appreciation of its importance in the context of managing cost and enhancing value in projects.

6.1 THE CONCEPT OF CASH FLOW

Simply defined, cash flow is the movement of funds in and out of a business, while cash flow management focuses on the timing of moving funds. This is often a matter of great importance to project managers, because even if a project is making good technical progress and is on schedule, it will be considered a financial failure if it runs out of money. Projects that suffer from poor cash flow ultimately incur additional costs and, possibly, significant delays as well. Sometimes, however, even borrowing additional money or stopping work until funds are received may not be viable options.

There are a variety of methods for funding projects, ranging from internal sources (the project is part of the corporate budget) to external means (use of government "seed" money or direct investment). The way in which projects are financed has a major bearing on the cash flow arrangement for each of the project participants.

Because the various participants in the project have different interests in its outcome, their attitudes toward it and the way in which they are paid can all be quite different. For example, an organization that sponsors a major capital development project may require funds from a variety of sources. This often results in the formation of a consortium of interested parties (typically companies or banks) who put up money to finance the creation of the asset, and later fund its operation. If the asset can be exploited successfully, the operating revenues eventually generate a sum greater than the capital, and operating costs and profit accrue to the consortium members. This model has served as the basic arrangement for financing numerous projects, including the Channel Tunnel. The typical cash flow for a sponsor of a capital project (i.e., one capable of commercial exploitation upon completion) is shown in Figure 6.1.

Notice in this figure that operational costs often start to accrue before the completion of the capital project, while revenue may not start to flow until well *after* completion. Furthermore, the capital cost may have a funding gap in the operational phase that must be bridged. Also, the breakeven point for the project may be significantly later than the project's start date; for example, it may require as much or more time to reach break even as it does to construct the capital asset itself.

The contractors who actually carry out the work of designing and building the asset are likely to be in a fundamentally different position than project sponsors. If they are not part of the sponsoring consortium, they will have no interest in the commercial exploitation of the asset; all they want is to be paid for their work as and when it is completed, and move on to another project when their assignments are complete. The arrangement for making claims for payment will normally be covered in the contract, and a variety of payment schemes have been devised, depending on the project's nature and circumstances. On the basis that work is paid for as it is performed, the contractor's cash flow will generally look like that shown in Figure 6.2.

Referring back to Figure 6.1, notice that the contractor's position has little or no impact on the eventual outcome of the project. Assuming that

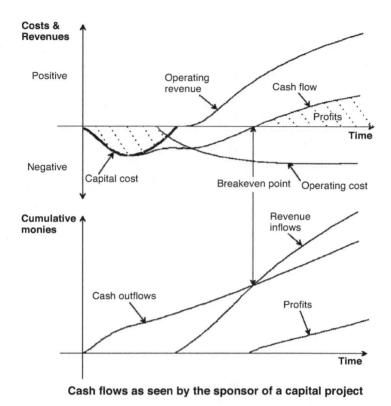

Cash flows as seen by the sponsor of a capital project

Figure 6.1 Cash flow curves for the sponsoring organization of a capital project[1]

work is done satisfactorily, a final profit will accrue at or near the end of the contract period, and sponsors will generally allow profits to be added into the claims for payment as they are made, even under cost-plus-fee terms. Even then, it is generally not possible to do the work without a cash float over most of the project's duration, as shown in Figure 6.2.

This float is required because claims are generally retrospective, i.e., for work done in the past, and there is always a delay between making the claim and actual payment. The extent to which the contractor is exposed will reflect both the agreed-upon contract conditions and how well he or she performs the work. With a simple cost-plus reimbursement arrangement, cash exposure may be limited to about 60 days' working capital, and may be even less if advance payments for "start-up costs" can be obtained. However, with milestone or stage payment plans (discussed later in this chapter), the exposure can be much greater if the work is delayed or not performed satisfactorily.

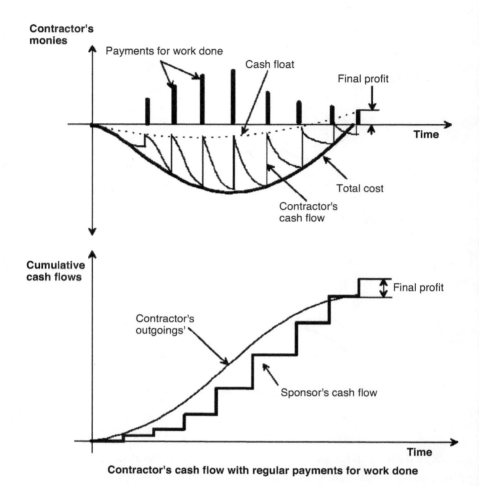

Figure 6.2 The contractor's cash flow profile involves much less cash exposure[2]

Note: Private venture projects for the development of new products and services put the developer in the same position as the sponsor of a capital project, although the departments within the company that do the work may, for accounting and control purposes, be treated as though they are contractors.

From both the sponsor's and the contractors' points of view, the importance of managing the cash situation is critical. Both parties need to know which expenditures are taking place as they happen. The sponsor must plan to have access to funds to pay the contractors, and must be aware of growing costs and progress being made on the project. The contractor must have systems in place that correctly gather costs as they are incurred and generate billing at the appropriate time. There should also be a credit control system that ensures payment is received within an acceptable timeframe.

Some of these aspects will be contained within a company's general accounting system and overall terms of trading. Even though the project manager may not have much influence or control over these issues, it is the responsibility of all project managers to ensure that appropriate systems are set up to gather relevant data and feed it to the general accounting and invoicing system.

Project managers themselves are not normally required to manage the actual cash position, unless the project takes up a large part of the company's resources. Instead, this is often done centrally, and many companies, particularly the large ones, will draw cash from a variety of sources. These can include concurrent projects, ongoing sales of production items or repeat services, loans and overdrafts, rights issues, etc. Funds from these diverse sources are normally pooled into the business.

6.2 CASH FLOW AND THE WORTH OF PROJECTS

The term **cash flow** is used to describe the net difference, at any point in time, between income (revenue) and project expenditures; negative cash flow is outgoing, while positive cash flow is income. Figures 6.1 and 6.2 show that at some point in the project life cycle, the revenue line crosses the expenditure curve. Beyond that point, the project will generate a profit.

In the case of a pure contractor with no interest in the operational phase of the project, a cash flow breakeven analysis may not be performed. If one is undertaken, it is far more likely to be done in a competitive tendering situation, for the purpose of a risk/reward assessment. However, with significant capital projects, which may take years to complete and have an operational or production life of twenty or more years, establishing the breakeven point is a fundamental aspect of the project investment analysis.

With major projects, where development costs are enormous, the investment decision is based on the anticipated number of sales and the period over which the product will be sold. For example, the development costs for Airbus A380 project may exceed $20 billion, and the breakeven number for airliners at the time of project launch is estimated to be over 100 units. In reality, however, the actual number required to make a true profit is likely to turn out to be much greater. Clearly, if sales are slower than ex~~~~~~~~~~~~~~~~~~~~~e ongoing costs of operating an airline pr~~~~~~~~~~~~~~~~~~point will rise rapidly and extend much further into the future than anticipated.

6.2.1 The Time Value of Money, and Techniques for Determining It

In the case of longer-term projects, the time at which profits start accruing can have a bearing on the worth of future earnings. This is expressed by an important financial concept called the **time value of money**. It states that, in general, money earned now is worth more than money earned at some time in the future, for two reasons:

1. Additional return could have been obtained if the money had been invested in the intervening period.
2. Purchasing power is reduced, due to inflation.

Furthermore, profits made much later in the future are more vulnerable to changes in circumstances than near-term earnings; for this reason, less value should be attached to them.

In the case of long-term projects, it has become normal practice to consider the time value of money by applying a **discount rate** to future cash flows. The discount rate used is based on the required or desired rate of return, and may also include the expected rate of inflation over the life of the project. Applying discounting rates tends to show very clearly that projects generating a profit early in their life cycle are preferred to projects that generate profits much later in the future.

When financially evaluating both capital and developmental projects, it has become common to employ the technique of **discounted cash flow (DCF)**. Discounted cash flow takes into account the timing at which expenditure and profits rise, and shows the effective rate of return on the total investment over the life of the project. When evaluating the worth of the project, the desired rate of return can be set at some boundary level. If the initial analysis and evaluation shows that this rate of return cannot be achieved, then the returns are likely to be too low and the project will not be worth pursuing. The method of calculating DCF is relatively straightforward, and can be easily adapted to spreadsheet methods.

An alternative approach is to determine the discount rate that would generate a zero return over the complete life of the project. This is termed the **internal rate of return (IRR).** A hurdle rate can be set for the IRR, and any project that cannot meet or exceed it is deemed to be not worth pursuing unless there are other important, nonfinancial factors.

Figure 6.3 shows the sensitivity of the breakeven point in terms of the time at which breakeven is achieved, from the point of view of both discount rates and project progress. While these are not the only factors that determine profitability, they do emphasize the point that shorter-term projects are likely to be less vulnerable than longer-term ones.

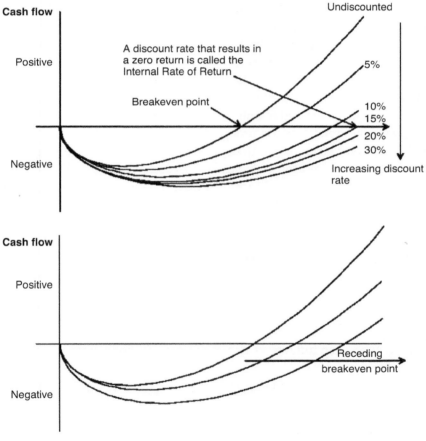

Effects on the worth of the project due to increasing discount rates and delays in achieving the breakeven point

Figure 6.3 Cash flow characteristics showing how a more pessimistic view of future discount rates and delays in achieving the breakeven point can dramatically reduce the net worth of the project[3]

6.2.2 Applying Discounting to Project Cash Flow

Completing the construction or development phase in the shortest reasonable time should always be a project management goal, unless there are very good reasons for a delay. For example, if the project organization determines that uncertainties in the labor markets or political arenas of host countries make it difficult to execute a project in the short term, it may choose to deliberately slow down the project until these uncertainties resolve themselves. When the discounting process is applied to project cash flow, it produces the **net present value (NPV)** of the proposed project at the given discount rate. (A project will have an NPV of zero at the IRR.) This is a popular method of establishing the worth of a project investment, with a positive NPV at the hurdle discount rate identifying a potentially viable project from a financial point of view.

In addition to the NPV method, the **discounted cash flow return (DCFR)** approach can be used. This approach is particularly suitable if capital is being borrowed to finance the project. With this method, besides the pure cash inflows to the project, interest is assumed on the negative cash balance at a rate equal to the applied discount rate. This inevitably produces a much greater cash demand than the simple DCF calculation, and is the equivalent of paying off a mortgage. In the early period, any cash inflows are absorbed in paying the interest, but as the project nears completion, the debt rapidly diminishes.

Regardless of the method used, the result for a given discount rate will be the same: a positive NPV will also result in a DCFR that is above the discount rate used in the NPV calculation. At the project's internal rate of return, both methods will generate a value of zero.

The formulas for calculating NPV and IRR are as follows:

1. NPV = (Cash inflows − Cash outflows) ∗ Discount rate
   where
   Discount rate = $1/(1 + k + r)^t$

   and
   k = the inflation rate
   r = the required or desired rate of return
   t = time period

2. IRR = the discount rate at which NPV is approximately equal to 0.

Cash Flow and the Worth of Projects

Example 1

Your project organization has to decide whether or not to invest in a project opportunity. The following information is available to you:

Initial cash outflow = $200,000 in the current year (year 0), and $100,000 in the next year
Cash inflows = $100,000 in year 1, $150,000 in year 2, $175,000 in year 3, and $75,000 in year 4
Required rate of return = 12%
Inflation rate = 4%

a. Calculate the NPV for this project.
b. Calculate the IRR for this project.

Solution

The discount factor and NPV for this example is given by

Discount factor = $1/(1+k+r)^t = 1/(1+0.04+0.12)^t$ for $t = 0, 1, 2, 3,$ and 4
NPV = Net flow * Discount factor

a. **NPV** (Table 6.1)
b. **IRR: Steps** (Tables 6.2.1, 6.2.2, and 6.2.3)

1. Assuming an inflation rate of 4 percent, try different rates of returns to calculate the NPV of the cash inflows.
2. Subtract the cash outlay from the total NPV of the inflows.
3. The rate at which the value from step 2 is close to zero is the IRR.

Table 6.1 Net present value calculations for the example project

Year	Cash inflows	Cash outflows	Net flow	Discount factor	NPV
0	0	$200,000	($200,000)	1.0000	(200,000)
1	$100,000	$100,000	0	0.8621	0
2	$150,000	0	$150,000	0.7432	111,474
3	$175,000	0	$175,000	0.6407	112,115
4	$ 75,000	0	$ 75,000	0.5523	41,422
TOTAL					**$65,011**

Table 6.2.1 Net present value calculations to determine IRR for the example project: Trial 1, $r = 12\%$

Year	Cash inflows	Discount factor	NPV
1	$100,000	0.8621	86,207
2	$150,000	0.7432	111,474
3	$175,000	0.6407	112,115
4	$ 75,000	0.5523	41,422
Total NPV of cash inflows			351,218
Total cash outlay			286,207
NPV difference			**$65,011**

Table 6.2.2 Net present value calculations to determine IRR for the example project: Trial 2, $r = 15\%$

Year	Cash inflows	Discount factor	NPV
1	$100,000	0.8403	84,034
2	$150,000	0.7062	105,925
3	$175,000	0.5934	103,848
4	$ 75,000	0.4987	37,400
Total NPV of cash inflows			331,207
Total cash outlay			284,034
NPV difference			**$47,173**

Continuing with the trials for $r > 12\%$, you will find that for $r = 24.7\%$, the returns are close to zero (about −$11). Hence, a close approximation of IRR for this example is 24.7 percent. If the firm required a return greater than 24.7 percent, the project will not be acceptable.

Table 6.2.3 Net present value calculations to determine IRR for the example project: Trial 2, $r = 24.7\%$

Year	Cash inflows	Discount factor	NPV
1	$100,000	0.7770	77,700
2	$150,000	0.6037	90,559
3	$175,000	0.4691	82,092
4	$ 75,000	0.3645	27,337
Total NPV of cash inflows			277,689
Total cash outlay			277,700
NPV difference			−$11

As a result of the above analysis, we can make some reasonable decisions on whether or not a project is worth the investment. There is an important caveat, however, when using net present value to evaluate projects. The longer the project's life or projected stream of revenues, the less precise this approach becomes. The reason for this is that we are making certain assumptions about critical issues, like the stability of the inflation rate over time. If inflation suddenly rose dramatically, in the midst of a 10-year repayment schedule, it would take longer to pay off the initial investment. Also, in uncertain financial or economic times, it can be risky to make investment decisions when discount rates may fluctuate.

Some drawbacks with using internal rate of return must also be taken into consideration. First, IRR is not the rate of return for the project. In fact, IRR equals the project's rate of return *only* when project-generated cash inflows can be reinvested in new projects at similar rates of return. If the firm can reinvest revenues only on lower-return projects, the "real" return on the project is something less than the calculated IRR. A couple of other problems with IRR make NPV a more robust determinant of project viability[4]:

- IRR and NPV calculations typically agree (that is, make the same investment recommendations) only when projects are independent of each other. If projects are not mutually exclusive, IRR and NPV may rank them differently. The reason is that NPV employs a weighted average cost of capital discount rate that reflects potential reinvestment, while IRR does not. Because of this distinction, NPV is generally preferred as a more realistic measure of investment opportunity.
- If cash flows are not normal, IRR may arrive at multiple, conflicting solutions, as is the case when net outflows follow a period of net cash inflows. For example, if it is necessary to invest in land reclamation or other incidental but significant expenses following the completion of plant construction, an IRR calculation may result in multiple return rates, only one of which is correct.

6.3 PAYMENT ARRANGEMENTS

Without a steady source of funds, the project will simply grind to a halt. The way to ensure continuous cash inflow depends on the nature of the project, and who the sponsor is. With externally sponsored projects, payment for work done is covered by the contractual arrangements set up at the time of project initiation. With internally funded projects, there will normally be some budget approval process by which funds are allocated. With the latter approach, the project manager is free to deploy resources on the project and use the funds, as long as the expenditures are within the approved budget limits.

Expenditures against budget will normally be monitored through the company's accounting system. These systems typically gather time bookings against the approved project code numbers (numbers often generated from the project's WBS coding) and then calculate costs using the appropriate charge rates. Brought-in materials, parts, and services are acquired through the purchasing system, which is separately related to the accounting system.

Accounting codes will also be generated for other expenditures, such as personnel expenses. At regular intervals (usually monthly), the accounting system should be able to generate figures for the amount spent in a given period against the various cost codes and cumulative totals. Project managers can then compare actual expenditures with approved budgets. If it appears that the budget limit is likely to be breached in the near term, then the project manager will normally be required to submit a request for a budget increase to the approving authority. How these requests are viewed will depend on many factors, including the importance attached to the project and the reasons for the cost overruns.

Outside contractors who undertake project work on behalf of a sponsor normally expect to receive payments as and when jobs are completed, and make a profit at the end, as illustrated in Figure 6.2. Payment arrangements can vary between projects, and will reflect the degree of risk that each party is willing to take, as well as the degree of control the sponsor wishes to retain.

6.3.1 Cost-reimbursable Arrangements

A cost-reimbursable arrangement is often the simplest form of contract, particularly if the work is ill defined at the outset. This is also called the "cost-plus-fee" or "cost-plus" contract. It is sometimes referred to as a "limit-of-liability" contract, because it is normal for the sponsor to set limits on the amount of funds that he or she is willing to allocate. When that limit is reached, the sponsor can increase the limit, in which case work can continue. However, if the limit is not increased, the contractor is free to stop work.

Under these conditions, contractors are normally allowed to bill for costs incurred as and when they complete related work. Billing is often done on a monthly basis, and normally an allowance for profit will be added to each monthly bill. From the contractor's viewpoint, this is a particularly good arrangement as there are no risks associated with either work content or getting paid promptly. Cash flow exposure is usually limited to little more than the operating costs incurred during the period allowed for bill paying—typically 30 or 60 days.

Although this may be an advantageous arrangement for the contractor, there are some risks for the sponsor. For example, work that has been claimed as completed may not actually have been done. Because of this potential, some sponsoring organizations, such as government agencies, demand the right to have an independent audit of project accounts. They may also demand the right to send a representative onto the contractor's premises to view work in progress and ensure that claims for payment are in accordance with the progress achieved. In other cases, certificates may be required from a quantity surveyor confirming that the claimed amount of work has been done. With a contract of any size, it would be remiss of a sponsor not to do this, yet organizations that fail to secure these rights at the contract negotiation stage can easily find themselves in this position.

The project manager's most important responsibilities concerning cost-reimbursable arrangements are

1. To generate a set of work packages that clearly specify the work to be done, and from which any particular work package can easily be identified. Work package codes that can be used as timesheet or purchase order codes are ideal for this purpose. When choosing a coding system, however, it is important to use a format that is compatible with the existing accounting system. While it is possible to use work package codes that do not conform to accounting codes, it is an additional complication that can lead to extra work and become a source of errors.
2. To ensure that all staff accurately account for time spent working on the project. All personnel should record their time manually on timesheets or through electronic input. It is important that the

correct booking codes are used, and project managers must ensure that responsible section heads enforce compliance to this rule. The same rule should also apply to codes attached to purchases.

3. To ensure prompt invoicing for work done. All work accomplished during the qualifying period should be accounted for, and the appropriate invoices should be prepared. While credit control associated with bill payment may not be the responsibility of the project manager, it should be brought to his or her attention if the sponsor is consistently late on settlements.

6.3.2 Payment Plans

Payment plans are a useful way of regulating cash-flow exposure for the contractor and, to a lesser extent, the sponsor. However, they do rely on a well-defined and clearly understood plan that is largely adhered to throughout the life of the project.

With some types of contracts, sponsors will wish to limit payments to the extent that only planned and completed work is paid for, given that the work can be well defined in advance. This approach is potentially applicable to either the cost-reimbursable or the **fixed-price (lump sum)** contract. The critical feature is that the work is clearly understood and delineated well in advance, and set out in a plan to which both parties have agreed.

With the simple fixed-price contract, arrangements can include looking across the complete project plan and defining a series of easily identifiable points of achievement, such as the completion of the first prototype or completion of all functional tests. These milestones mark the progress of the project. Then, the total value of the project up to each milestone is calculated, which in turn generates the payment plan, based on the timing of the milestones.

Because there can be periods of several months between each milestone, contractors are often allowed to bill monthly for work as it is completed, up to some fixed percentage (e.g., 75 percent) of the value of the work related to that milestone. Beyond that, no further billing is allowed until both parties have agreed that the milestone has been met. This arrangement is advantageous to both sides, because it allows claims to be made as work proceeds, which aids the contractor, while ensuring that only completed work is paid for.

Another advantage is that the sponsor can clearly see the total amount of money he or she is expected to pay throughout the life of the project, which can help when arranging financing. In addition, the prospect of delayed payment due to lack of progress is an incentive for the contractor to meet scheduled due dates.

Fixed-price milestone payment plans can also be used in a cost-reimbursable arrangement, provided overall budgets can be agreed upon between the contractor and the sponsor. Claims can be made as before, up to a set percentage of the budget value attached to the milestone, but once the milestone has been met, the contractor can submit a balancing invoice for the actual costs incurred.

The drawback with milestone payments is that if the plan is subject to frequent revisions, it has to be completely rebudgeted and rescheduled each time there is a major change. This, in turn, leads to revisions in the time and value attached to the revised milestones. Another obvious drawback with fixed-price contracts is that they assume it is possible to be reasonably certain of the stages in project development and overall costs. When unanticipated problems arise, a fixed-price contract may suddenly make the project very unattractive for the contractor, resulting in demands to renegotiate or even terminate the project midstream.

There is a derivative of the milestone approach that overcomes, in part, the need to redefine the cost and timing of the work each time there is a revision to the plan. With a **stage payment** arrangement, the project is divided into a series of blocks of work, or "stages," each with a budget attached. Usually, these stages are directly related to the major areas of work in the WBS.

Unlike milestones, which are discrete events that occur in sequence, it is possible for work on several stages to be performed concurrently. If the stages are relatively large, billing can be done monthly up to some fixed percentage, as is done in the milestone plan, Beyond that point, billing is not allowed until the agreed-upon stage is completed. In a big project involving a large number of stages, where work on a substantial number is performed concurrently, more extensive administration is needed to ensure correct monthly billing. This is because each stage is likely to be at a different level of completion, and will have a different value.

The advantage of the stage payment arrangement is that when a project is reconfigured, it is likely that only a proportion of the work will be changed, and only those stages that are directly affected by the change will need revision. In some cases, only the planned completion date will change—the basic work content will remain unaffected. Stage payment arrangements also allow contractors to be paid promptly if they get ahead of schedule in some areas. However, this practice should not be allowed too often, because it can result in meaningless work if changes are instigated after tasks have been completed ahead of when they were planned.

6.3.3 Claims and Variations

It is the rare project that is completed without changes. Not only are changes expected, but they are also taken into account by most contracts. In extreme cases, sponsors have attempted to employ "all-risk, ceiling price" contracts, which state quite clearly that whatever may transpire on the project, the contractor is expected to complete the work and the sponsor will not pay any more than the agreed-upon ceiling price. This type of contract was used on aspects of the Channel Tunnel work in Europe, and led to a series of disputes between sponsor Eurotunnel and contractor Transmanche Link, particularly because the sponsor instigated some of the changes.

Although this practice seems unreasonable from the contractors' point of view, many massive capital projects could never be justified at the outset without some limitations on the cost of development. The fact that the Channel Tunnel's construction cost significantly exceeded its original estimate, for example, left the entire project in a terrible financial state.

With more normal contractual arrangements, the cost of work over and above that covered in the contract can normally be recovered in two ways: **claims** and **variations**. Employing claims and variations is a useful way of limiting cash exposure on the part of contractors in a situation that is subject to change. These two terms are sometimes used interchangeably, but in fact refer to two very different things.

A claim is generally a retrospective demand for compensation for costs properly incurred. Claims can be made whenever additional expenditure has been incurred for labor or expenses that are not in the agreed-upon scope of work. Sponsors typically wish to be informed as soon as it is determined that additional work or expenditure is needed, so that they can comment on what is being done, make suggestions that might limit costs, and also make the necessary provisions for cash.

In cases where the additional expenditures were necessary for the work to be completed, a claim can be raised to cover all the costs. Generally, these claims are made at cost, plus the normal profit. Claims should be submitted as soon as the additional expenditures are incurred, to properly account for what has been done, and should be separately identified from any routine invoices. If the claims submitted are challenged by the sponsor, particularly if they seem excessive, then such disagreements should be resolved through negotiations. It is important to note that prices contained in the scope of work cannot be varied if the work is performed under a fixed-price arrangement, even if the contractor overspends.

Unlike claims, variations cover work that will take place in the future. Variations are inevitable on most projects, and can be a major source of contention when firms underbid to win work and then look for ways to make up the difference during development. There is no doubt that in many competitive situations, there is full knowledge on the part of the contractor that the quoted price cannot possibly cover the anticipated work. In these situations, variations to the scope of work are introduced as the work proceeds, and, if carefully managed and presented, can change the contractor's potential loss into a profitable state.

While these practices by contractors certainly constitute exploitation of the customer's lack of knowledge, they are also a defensive reaction to sponsors who seek to offload their own risks by demanding fixed-price bids in competitive situations. This practice of deliberate underbidding has become so common, some regard it as part of the normal process of attracting and maintaining business.

Variations can take a variety of forms, depending on the project and contract terms. They can arise when sponsors change their demands, when contractors make suggestions, or when there is acceptance of changed circumstances. Depending on the nature of the change, negotiations may be required for agreement on the content of the revised work and the price to be paid.

Once agreed upon, the variation becomes a contractual amendment to the original scope of work and is usually contained in a "variation order," or a "contract change note" document. Acceptance of the variation order allows the contractor to proceed with the new work and present invoices for payment in the normal way. Not all variation orders cover increases to the project. Occasionally, aspects of the project work are no longer deemed necessary, due to changed circumstances, and a variation order is raised to delete them from the scope of work.

6.3.4 Cost Variation Due to Inflation and Exchange Rate Fluctuation

Despite the fact that governments and forecasting institutions have spent huge sums on developing economic models, the ability to accurately forecast inflation rates over the medium to long term remains extremely difficult. The effect of inflation on program cost appears in two ways:

1. Inflation over and above that allowed for in the estimate, which generates additional cost over which the contractor has no control
2. Inflation for projects subject to slippage, where not only is there more work than expected, but also more is being done later than planned, and thus at a higher-than-expected cost

This situation clearly has an adverse impact on the real value of payments made under fixed-price conditions. To eliminate this risk, it has become common practice to include a clause in the contract that allows the sum of money paid at any point in the project to be varied according to the prevailing rate of inflation. This was particularly the case during the 1970s and 1980s, when inflation rates were relatively high.

Contracts of this type have been viewed as fair for all concerned parties, because they remove a contentious element of risk: in terms of purchasing power, the relative value of the contract remains fixed. Neither party is expected to shoulder the total burden of cost increases if inflation turns out to be higher than expected, or make an excess profit if it is lower. Necessary adjustments to price are made by applying formulas that use published index number series.

When projects involve participants outside the host country, currency fluctuations become another aspect of value that is subject to variation. The methods for dealing with this can involve such things as agreeing with the purchaser that all invoices will be in one currency and paid in that currency. For example, if it is agreed upon that all invoices will be issued in U.S. dollars and all payments will be in dollars, all of the exchange rate risks lie with the non-U.S. purchaser. If, however, a U.S. firm agrees to issue all of its invoices in euros and be paid in euros (as would be the case with a U.S.-flagged carrier purchasing Airbus aircraft for its fleet), the non-European party assumes the currency exchange rate risk.

Other methods to counter currency exchange rate movements include forward-buying the currency at a fixed exchange rate, and agreements to revalue the contract work at fixed times in the future, based on the prevalent exchange rates at those times.

6.3.5 Price Incentives

Occasionally, situations arise where there is a possibility that cost savings can be made during the course of the project. This can happen when uncertainties are so great that it is impossible to agree to a fixed-price contract without including excessive contingencies. Sponsors naturally want to avoid this option, and may instead propose a contract that includes an inducement for the contractor to finish the project at a price lower than the fixed amount. This inducement is in the form of increased profit, even though the total contract value may be less, and means that the customer is sharing a proportion of the cost savings with the contractor. Inducements can be in the form of a price adjustment formula that relates to the final profit, or a bonus payable if agreed-upon cost thresholds are met. Bonuses can also be paid for early completion.

It is normal to fix, at the outset, both the maximum price (**the ceiling price**) and the assumed minimum level of profit. Below the ceiling price, a lower figure is agreed upon as the **target price**, which contains a higher level of profits that the contractor feels he or she should be able to achieve. As the contract proceeds, claims are made in stages, according to a payment plan and observed progress. Profit at the low level can also be included in these claims. At the end of the contract, a formula is applied to the value of the claims, and if the total value is below the ceiling price an additional payment is made.

Incentive contracts may be devised that also allow for price adjustments due to inflation. If a project is expected to last five or more years, and if the sponsor wishes to introduce an incentive element, an escalation clause may be the only way it can be done. However, situations involving incentive contracts can become quite complex. Because of this, it is important to emphasize that these situations should be carefully examined before applying a formula to the value of the claims. Issues to be examined should include cost escalation due to inflation, as well as nonconformance with performance specification and project schedule. Examples of these types of price adjustments include

- *Contract price adjustments due to inflation*—These adjustments follow the employment of a simple price-modifier value, based on

changes in the inflation rate that imply some medium- to long-term stability.
- *Price adjustments due to schedule variations*—These adjustments take into consideration schedule compliance through both early completion and significant delays.
- *Price adjustments due to performance variations*—These adjustments assess project functionality and lead to contract increases based on superior performance, or penalties for lack of compliance.
- *Price adjustments due to cost variations*—These adjustments occur in cases of either significantly lower or higher costs than the contracted amount.

6.3.6   Retentions

Despite the fact that work may appear to be completed satisfactorily, there may be reasons to withhold an amount of money until final settlement is made with the contractor. This practice is called **retention**. Usually, retentions are used to ensure that if any defects are discovered in the project work, the contractor will properly rectify them. They can also be used to cover any accounting irregularities that may be discovered between pricing and invoicing. Retention is a separate arrangement from either a warranty for the product or liquidated damages for the effects of nonperformance. Retentions usually involve a small percentage (10 percent or less) of the contract price that is withheld for a short period (e.g., three months). Contractors, however, may object to the inclusion of significant retentions in their contracts.[5]

From a theoretical perspective, it can be argued that there is no direct correlation between cash flow and project profitability, because projects with severe cash flow problems may ultimately turn out to be profitable. In reality, however, there is a strong correlation between the two. Consequently, project managers can use the following guidelines to improve cash flow[6]:

1. *Build and maintain strong relationships*—Because people both inside and outside the project organization manage cash flow processes, it is vital that the project manager build strong relationships with all parties involved.
2. *Create a flowchart of each process that impacts cash flow*—This type of flowchart can help to visually identify the various

relationships in the process, redundancies, and potential ways to improve the process.
3. *Redesign the billing process*—The greatest opportunities for improving cash flow lie in this area. Ways to improve the process include creating a cross-reference between the billing items and the budget, or redesigning the billing format so that it is easy for customers to incorporate.
4. *Design suppliers and subcontractor pay scales and integrate them into your format*
5. *Streamline the submittal process*—For projects involving submittals, create a detailed submittal log and integrate the submittal log dates with schedules to optimize cash flow. Assign responsibilities for all line items, with built-in financial incentives and sanctions in pay schedules to accelerate the process.
6. *Reduce change order processing time*—Change orders have a negative impact on cash flow due to slow approval processes and eventual payment. To minimize unnecessary delays, manage any change request as effectively as possible.
7. *Develop knowledge of compliance and administrative documents*—Even though these documents are under the purview of the contracts, accounts receivable, and payroll departments, the project manager should have an adequate knowledge of them in order to ensure that associated processes occur in a timely manner.
8. *Manage the project closeout process*—More often than not, a significant portion of monies due is still held by the customer in the form of final payment, retention, or both. Therefore, how quickly these monies are released is dictated by how well the closeout process is managed. It is important to start this process as early as possible by communicating with all project stakeholders to learn what steps need to be taken to expeditiously close out the project.
9. *Get everyone involved in the cash flow game*—Managing and improving project cash flow is serious business, so involve all stakeholders (accounting, contracts, accounts payable, accounts receivable, etc.) who can improve the cash flow cycle.

Managing cash flow is a surprisingly complex component of the project development cycle. As this chapter has demonstrated, the idea that it involves cutting checks and keeping track of payments enforces a far too simplistic perspective that routinely gets project managers and their projects into deep trouble. Indeed, the number of projects that have had to be canceled due to excessive overruns and poor cash flow is huge and growing. This chapter has offered some suggestions and avenues for developing a clear appreciation of the importance of cash flow and some means to better manage the process, ensuring that the project maintains a healthy balance sheet throughout its development.

REFERENCES

1. UMIST Module 4 workbook, p. 4.3.2.
2. UMIST Module 4 workbook, p. 4.3.3.
3. UMIST Module 4 workbook, p. 4.3.7.
4. Keown, A. J., Scott, D. F., Jr., Martin, J. D., and Petty, J. W. (2010) *Foundations of Finance: Logic and Practice of Financial Management*, 6th ed. Upper Saddle River, NJ: Prentice Hall.
5. UMIST Module 4 workbook, pp. 4.3.8 to 4.3.26.
6. Brown, D. (n.d.) Ten ways to improve cash flow. http://www.ecmweb.com/construction/electric top ten ways/index.html

KEY TERMS

Cash flow
Time value of money
Discount rate
Discounted cash flow (DCF)
Internal rate of return (IRR)
Net present value (NPV)
Discounted cash flow return (DCFR)

Fixed-price
Lump sum
Stage payment
Claims and variations
Ceiling price
Target price
Retention

Chapter 7

Financial Management in Projects

LEARNING OBJECTIVES

- Describe financial management as applied to projects.
- Illustrate types and sources of finance.
- Demonstrate how to control financial risk.

Project funding and financial management have a significant impact on project cost, cash flow, and more importantly, success. And yet, very few project managers have even a rudimentary understanding of this important element of overall project strategy. It is important to recognize that the means a firm uses to finance its projects can have a huge impact on their ability to successfully control costs, manage cash flow, and maintain an acceptably positive degree of value for the project. Put another way, when we make errors in selecting the manner in which we choose to finance a project investment, it can have a great impact on the ultimate worth of that project. Thus, it is necessary, within the scope of this book, to make some general mention of the role of financial management in successful projects.

In this chapter, we discuss project funding and financial management, including some of the most important concepts and key issues. We'll differentiate between financing a project and project finance, and explore the different types and sources of finance. Then, we'll cover the various steps involved in the process of financial management of projects, and conclude with possible financial risks.

7.1 FINANCING OF PROJECTS VERSUS PROJECT FINANCE

At the outset, it is important to understand the difference between the terms **financing of projects** and **project finance**. Typically, most projects are paid for by the parent organization, either from revenues or capital expenditures. If the parent organization borrows the finances needed, the money must be repaid, regardless of project success or failure. We refer to this type of financial arrangement as **financing of projects**.

A few large projects are treated as distinct entities and are based on unsecured or limited recourse financing. This type of financing is unsecured in the sense that the money borrowed is secured only against the project's assets and its revenue stream. Because the investment does not show up on the company's balance sheet, it is also referred to as off-balance sheet financing. If the project fails, the lender has no recourse to recover the investment.

In limited recourse financing, the parent company has some equity in the project, which will show up on the company's balance sheet. This type of nonrecourse or limited-recourse financing is referred to as **project finance**.

7.2 PRINCIPLES OF FINANCING PROJECTS

There are several important principles to consider in the financial management of projects. First, the money required to finance a project is the single largest component of project cost. While costs associated with materials, construction, working capital, etc. can amount to a significant portion of the total, no other cost component is as large. In fact, when one takes into consideration the interest paid to debtors and returns to equity holders, the total cost of the financial package can be twice as much as the next-largest cost component. Unfortunately, many project managers are unaware of this cost impact and make the mistake of minimizing capital costs, instead of minimizing the total cost (including the cost of financing).

Second, when selecting projects to be commissioned, only those projects that have a complete financial package in place will get the nod. For projects financed by the parent organization, a financial package is typically not put forth until design work is completed, because the design package is considered part of the investment appraisal process. However, in the case of projects that involve nonrecourse financing, design work often does not begin until financing is obtained.

Third, financial planning should begin at the feasibility stage, and should be an integral part of the overall project strategy. Unfortunately, in most project scenarios, financial planning often begins after all other project features, including the technical solution, have been determined. In the most successful projects, however, financial planning is a key feature of project strategy from the very beginning. In addition, involving financiers as early as possible in the planning stage will eliminate high-risk options and pave the way for project funding. The goal is to minimize life-cycle costs, including finance charges incurred, along with the cost impact of technical solutions adopted.

Fourth, projects are complex by nature, and financial planning further aggravates this complexity. This is particularly true for projects that involve nonrecourse or limited-recourse financing (project finance), because they are typically mammoth projects that can take several years to complete, and may not deliver a return on investment for several decades. These types of projects usually involve a single product or purpose, cut across national boundaries, are undertaken by several organizations, and involve a multitude of contractors and subcontractors and tiers of suppliers.

Financial planning for such projects requires that sources of finance be identified before work begins on technical specifications for the project or its equipment. In these cases, the financial package often acts as a constraint on technical scope, and may require compromise. Furthermore, because these projects involve no-recourse or limited-recourse financing, lenders may prefer low-risk technical solutions. If they do prefer high-risk solutions, they are likely to charge higher interest rates that will significantly increase the project's financial cost.

7.3 TYPES AND SOURCES OF FINANCE

There are two main types of financing: equity and debt. Equity financing refers to the money subscribed by investors and shareholders, whose returns on investment are in the form of dividends and capital growth equivalent to the value of their equity in the project organization. However, the project organization can make dividend disbursements only after the interest and scheduled loan repayment obligations have been met. If the project is successful, the returns on investment for equity holders can be substantial; if it fails, equity providers may receive no returns. Consequently, equity holders may demand higher returns than other debtors.

Debt financing refers to money borrowed from a number of sources, including banks. This debt involves periodic repayments of the debt and interest, based on agreed-upon schedules. The money borrowed through this type of financing arrangement has to be repaid first, before the repayment of other types of finances. This type of debt is also sometimes known as **senior debt**.

The providers of senior debt have the first claim to the project organization's assets should the project fail and the company goes into liquidation. Also, debts can be secured or unsecured. In the case of secured debt, the money lent to the project organization is secured against its assets and usually carries a lower interest rate when compared to unsecured debt. This money has to be repaid whether or not the project is successful.

Unsecured debt is secured only by the assets and revenues of the specific project for which the money was lent, and lenders have no recourse to recovering money from the assets of the project organization should the project fail. Due to the higher risk assumed by the lenders, unsecured debt carries higher interest rates than secured debt.

Another type of debt financing involves loans from the project organization's equity holders. This type of debt, called **Mezzanine debt**, involves a schedule of loan repayments and interest payments at a predetermined rate. Mezzanine debt, however, is considered secondary to the senior debt discussed earlier, and all loan and interest repayments on mezzanine debt can be made only after financial obligations to the lenders of the

---

Case Study: Eurotunnel: Problems with Cash Flow
"Eurotunnel, the construction of the channel tunnel between England and France, was an example of a limited-recourse financed project, involving a mixture of debt and equity. Equity was provided by the project promoters, mainly by the contractors involved in the construction of the tunnel, and by private investors. But the vast majority of the finance (in excess of 80 percent) was through loans provided by banks.

During the construction of the tunnel, interest on the loans was added to the debt. During the early stages of the project, the equity holders received some returns on their investment in that the share price steadily rose as the commissioning date, and hence the expected return, became closer. However, the early revenue streams from the project were not enough to cover the schedule of debt and interest payments to the banks.

The shares now had no value and the equity investors lost their investment, and the project effectively became a nonrecourse financed project. There was no point, however, in the banks foreclosing on their loans, because the project's main asset (the hole in the ground) had no value if it was not being used. Hence, the banks converted much of their loans to equity, and are now receiving their returns over the life of the project in the form of dividend payments."[1]

senior debt have been fulfilled. Consequently, mezzanine debt involves higher risk and correspondingly higher interest rates.

7.4 SOURCES OF FINANCE

There are both conventional and unconventional sources of finance. The conventional sources include the company's shareholders, banks, suppliers, export credit, buyer and seller credit, and international investment institutions such as World Bank and development banks such as the Asian Development Bank. Unconventional sources of finance include leasing assets, counter trade, forfeiting, switch trade, and debt/equity swapping.[2]

- *Leasing assets*—rather than purchasing a project asset, the project organization leases it from a third party, who receives a return in the form of rental for the asset.
- *Counter trade*—The seller accepts goods or services in lieu of cash. For example, it is quite common in poorly developed or third world countries to "pay" for a portion of a large project with commodities, which the project developer then must sell in order to raise cash.
- *Forfeiting*—finance is made available through the sale of financial instruments due to mature at some time in the future. Finance is then provided by trading in these assets in the futures market.
- *Switch trade*—This process makes use of a credit surplus between two parties to finance a relationship with a third party. For example, if country A has a credit surplus with country B, exports from C to A can be financed with payments from B to C.
- *Debt/equity swapping*—A multinational company may buy a host country's debt at a discount. This is redeemed in local currency at favorable exchange rates and is used to set up local companies. These are used by the multinational company to transfer technology, generate foreign exchange, and create employment in the host country.

7.5 COST OF FINANCING

The cost associated with borrowing money depends on the particular form of capital borrowed. For example,

- The **cost of equity** is the dividends paid to shareholders plus any estimate of the equity's capital growth. The cost of equity is usually calculated using the capital asset pricing model (CAPM). (More can be learned about this model in any finance textbook.)

- The **cost of debt** is the cost of debt financing, or the interest paid on the money borrowed. While the cost of equity is payable out of untaxed income, the cost of debt is payable out of taxed income.
- The **cost of capital** is the average cost of various forms of finance used by the project organization; specifically, it is the weighted average of the cost of the different types of capital borrowed. For example, if the project is financed through both debt and equity, then

$$Cost\ of\ capital = Ratio\ of\ equity * Cost\ of\ equity$$
$$+ Ratio\ of\ debt * Cost\ of\ debt$$

7.6 PROJECT FINANCE

As described earlier, project finance refers to unsecured, nonrecourse, off-balance sheet financing of an individual project. Usually, these types of projects are large infrastructure projects undertaken by a single or group of private organizations on behalf of a government.

The contractual arrangements for these types of projects can take a number of different forms. For example, in a "build–own–operate– transfer" type of arrangement, the private sector company that constructs the project owns and operates it for a period of time. During this concession period, the revenues generated are used to repay the loan and realize some profits. Then, project assets are handed over to the government.

A second type of contractual arrangement, "build–own–operate," involves projects that have no residual value after the concession period, such as the construction of a power station. In these cases, the project assets are not handed over to the government. Other types of contractual arrangements include build–operate–transfer, design–build–finance–operate, build–lease–transfer, and build–transfer–operate.

The typical structure of project finance is shown in Figure 7.1. At the center of this structure is the project organization, or in the case of joint venture projects, group of project organizations. These project stakeholders are sometimes known as special purpose or project vehicle (SPV). The company that operates the project is one of the partners.

The financing in these types of projects is a combination of debt and equity. The debt-to-equity ratio for the project is typically high, with a ratio of 4:1 not uncommon. Because of the high level of debt involved, the return to project sponsors is also high. In addition, the providers of

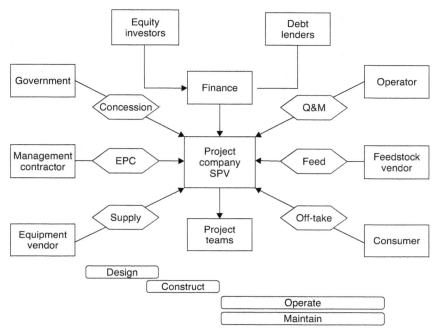

Figure 7.1   Project finance structure[3]

debt for these projects, usually banking institutions, require some equity investment by project partners to ensure their commitment and motivation to see the project through to a successful completion. Otherwise, the money invested by project stakeholders will be lost, and cannot be recovered.

Project finance involves two main types of contracts: concession agreements with the government, and off-take contracts with consumers. Both of these types are illustrated in Figure 7.1. In cases where the consumer is the government, the off-take contract is with the government. In these contracts, the revenue generated from the project is sometimes held in an escrow account that is controlled by debt providers. The money in the escrow account is first used to make the scheduled loan and interest payment to lenders. Any surplus is paid to the SPV to cover their costs, as well as disbursements to equity partners.

In addition to these main contracts, other ancillary contracts include engineering procurement and construction contracts with the managing contractor, equipment supply contracts with materials suppliers, operation and maintenance agreements with the operator, and so on.

The payments on these contractual obligations can take the form of direct or deferred payments, converted to loans or equity. If they are converted to loans, they are subordinate to bank loans.

7.7 THE PROCESS OF PROJECT FINANCIAL MANAGEMENT

The process of financial management involves five major steps: conducting feasibility studies, planning project finance, arranging the financial package, controlling the financial package, and managing financial risk.

7.7.1 Conducting Feasibility Studies

Feasibility studies allow the project organization to consider not only the various alternatives that maximize project value, but also the financial implications of these alternatives. Project alternatives are evaluated using financial appraisal techniques such as net present value analysis (NPV) and internal rate of return (IRR).

During this step, the financial objectives of the project are established. While these objectives differ between various project stakeholders, the main concerns of the project sponsor are:

- Raising necessary funds for the project at appropriate times, and in appropriate currencies
- Minimizing project cost and maximizing revenue
- Appropriate risk-sharing among all project stakeholders, including financiers
- Establishing adequate control and flexibility, including rescheduling loan and interest repayments if conditions warrant
- The ability to pay dividends to equity holders

7.7.2 Planning the Project Finance

In this stage of the financial management process, several key issues need to be considered. First, the total cost of the project and the amount of money to be borrowed have been identified from the feasibility studies. In addition, the timing of cash inflows and outflows to the project should be identified, including the time at which money needs to be borrowed.

Factors such as working capital requirements and the rate of inflation should also be taken into account.

Second, project sponsors should plan for an appropriate debt-to equity ratio that will enable them to optimize profits from the project while meeting financial obligations to lenders. Third, all possible sources of finance should be considered and identified, including who will be partners in the SPV, who will provide the required debt and/or equity, and whether sources of finance will be domestic, international, or a mixture of both.

Finally, after identifying the various sources of finance to the project, it is important to reanalyze cash flows to see if they are satisfactory, and, if possible, renegotiate with lenders to obtain more favorable interest rates.

7.7.3 Arranging the Financial Package

In this step, the financial package is put together by the project sponsor with the help of the main provider of debt (senior debt). This involves:

- Raising additional equity with the help of a merchant banker
- Identifying additional project sponsors and providers of subordinate debt
- Raising additional money through currency and bond markets
- Negotiating buyer/supplier credit and arranging insurance
- Exploring and negotiating other unconventional financial arrangements, such as leasing

7.7.4 Controlling the Financial Package

Once the project gets underway, finances should be carefully monitored and managed. During the project execution stage, actual expenditures should be compared against the preestablished plan, and corrective actions should be taken to eliminate or minimize deviations. At this juncture, it is also worthwhile to generate a forecast of cost to completion, so that additional finances can be arranged, if necessary. During the commissioning stage, initial operating and maintenance costs and initial sales revenues need to be monitored to ensure that financial objectives are being achieved according to plan. If not, corrective actions should be taken to eliminate deviations. During the operation stage, it is important to ensure that project assets are fully utilized. It is also important at this stage to keep lenders fully informed of project progress, so that there are no unpleasant surprises in the future.

7.7.5 Controlling Financial Risk

The various types of financial risks that can be encountered in a project are provided in Table 7.1. The project sponsor should carefully monitor the project for these risks, and should have a formal financial risk management procedure in place.

Clearly, some components of financial risk are easier to manage than others; for example, it is impossible to do more than be aware of inflation or interest rate risks, as they are elements of the macroeconomy. Nevertheless, it is critical to consider their potential impact when developing a financial risk strategy. For example, large oil companies devote considerable time and resources to remaining current on the political situation within countries where they do business. Recently, Venezuela announced their intention of nationalizing all oil development assets of foreign companies. This action leaves these firms with limited options. They can immediately abandon multibillion dollar plant and equipment assets or attempt to negotiate the best possible revenue- and profit-sharing arrangements they can hope to achieve with the current government. Thus, even though the oil companies work to determine and manage financial risks, they are still subject to the vagaries of political and social upheavals in countries where they have a strong vested stake.

Because some of these financial risks reside beyond the realm of effective management, it is critical that project organizations do manage the

Table 7.1 Potential financial risks in projects

Type of risk	*Examples*
Macroeconomic risks	Inflation, interest rates, currency and exchange rate fluctuations
Political risks	Country risks, legislation and regulations, change of government
Commercial risks	Viability and feasibility, cost and schedule completion, performance and operation, revenue availability, reliability, and maintainability
Contractual risks	Management, equipment supply, feedstock supply, concession and licenses, and sales agreements

risks over which they retain some element of control, such as commercial and contractual risks. Contractual risks demand that the project organization clearly specify financial obligations and performance requirements as a key element in any contract into which they enter with third parties. It is beyond the scope of this book to discuss in detail legal aspects of the management of projects but it bears mentioning that the best protection any organization can have from a variety of claims, penalties, and other contractual risks lies in careful understanding of the legal relationship between themselves and other third-party organizations.

Commercial risks are also an area in which our companies can make strong efforts to minimize negative impacts, depending on our understanding of the financial and performance risks to which we are exposed. In recent years, for example, a common method for attempting to control and minimize project commercial risks involves the use of options models as a means to determine the optimal time and circumstances in which to invest in a new project opportunity.

7.7.6 Options Models[4]

In Chapter 6 we discussed the concepts of the time value of money and the use of net present value (NPV) and internal rate of return (IRR) methodologies to assess the viability of a project opportunity. One key drawback shared by both the NPV and IRR approaches lies in their ignorance of an important assumption (i.e., the money invested in a project may be non-recoverable). In other words, many project investments may involve the decision to allocate money that cannot be recouped. For example, a firm building a power plant in a remote, poorly developed country may find it impossible to make a positive return on their investment. Further, they may find it impossible to find a buyer for the plant should they choose to abandon their investment. Therefore, organizations facing project choices that involve irrecoverable investments should determine whether[5]:

- The firm has the flexibility to postpone the project
- Information may become available later that will help the company determine whether or not the project is worth undertaking.

Consider an example: Assume that a construction firm is considering whether or not to upgrade an existing chemical plant. The initial cost of the construction upgrade is $5,000,000, which is estimated to generate cash flows of $1 million per year indefinitely. The company requires a 10% return on its investment. Assume that the chemical plant can be upgraded in one year and begin earning revenues the following year. The best forecast assumes that the organization will earn $1 million per year off this investment, but should adverse economic and political conditions prevail, the probability of this amount drops to 40%, with a 60% probability that the investment will yield only $200,000 per year. We can first calculate the net present value (NPV) of the investment as follows:

$$\text{Cash flows} = 0.4(\$1 \text{ million}) + 0.6(\$200,000) = \$520,000$$

$$\text{NPV} = -\$5,000,000 + \$520,000/0.1$$

$$= -\$5,000,000 + \$5,200,000$$

$$= \$200,000$$

According to the calculation above, the company should undertake the chemical plant construction project. However, if we ignore the possibility that by waiting a year, we may have a better sense of the political or economic climate within the host country, the firm is neglecting important information that could be useful in rendering the decision of whether or not to invest in the plant. Within a year, for example, we may be able to better determine whether the political situation in the country has swung in a positive or more negative direction, suggesting that the project would be too risky to undertake. For example, suppose that by waiting a year, the company will be able to ascertain that the investment will have a 50% likelihood of paying off at the higher value. If the earnings from the investment rise, the NPV for the project is now:

$$\text{NPV} = 0.5 \times [-\$5,000,000 + \$1,000,000/0.1]$$

$$= 0.5 \times [-\$5,000,000 + \$10,000,000]$$

$$= 0.5 \times \$5,000,000$$

$$= \$2,500,000$$

The implications of these two net present values are interesting. If the company has the opportunity to invest in the project, and has no other alternative with an NPV of $200,000, it should do so. On the other

hand, by w[...]   [...]e project rises dramatically, to
$2.5 millio[...]   [...]mpany's option and willingness
to wait the additional year) is worth an additional $2.3 million dollars.
Put another way, if the organization were attempting to calculate the
value of the option of waiting one year to undertake the project, they
should be willing to pay an additional $2.3 million dollars for a contract
that has this flexibility over one that does not.

Ultimately, it is critical to understand that projects require the effective support of an organization's financial management team to improve the likelihood of success. Securing sources of funding and minimizing exposure to risk are necessary components of successful project management. Done correctly, it can set the stage for a project to dramatically improve its cost management, cash flow, and value.

REFERENCES

1. Turner, R. (2004) The financing of projects. In: P. Morris and J. Pinto (Eds.), *The Wiley Guide to Managing Projects*. Hoboken, NJ: Wiley, pp. 344–345.
2. Turner, R. (2004) *Ibid.*; Dingle, J., and Jashapara, A. (1995) Raising project finance. In: R., Turner, (Ed.), *The Commercial Project Manager*. London: McGraw-Hill.
3. Turner, R. (2004) *Op cit.*, p. 352.
4. Pinto, J. K. (2009) *Project Management: Achieving Competitive Advantage*. Upper Saddle River, NJ: Prentice-Hall.
5. Dixit, A. K., and Pindyck, R. S. (1994) *Investment Under Uncertainty*, Princeton, NJ: Princeton University Press; Huchzermeier, W., and Loch, C. H. (2001) Project management under risk: using the real options approach to evaluate the flexibility in R&D, *Management Science*, 47(1): 85–101.

KEY TERMS

Financing of projects
Project finance
Senior debt
Mezzanine debt
Cost of equity
Cost of debt

Cost of capital
Leasing assets
Counter trade
Forfeiting
Switch trade
Debt/equity swapping

Chapter 8

Value Management

LEARNING OBJECTIVES

- Identify the five concepts that are necessary for enhanced project value.
- Describe the Value Management process.
- Discover how VM and Risk Management interrelate.

When an organization formulates new strategy, one of its key considerations is its ability to exploit opportunities in the external or internal environment. This includes initiating projects that enable the organization to achieve some well-defined goals or objectives—in other words, projects that add value.

This chapter offers a systematic means by which organizations can develop and apply an effective value management process. It begins with an exploration of the concepts of **value**, and particularly **project value**, including how it is derived and applied. Then, the focus shifts to the concepts, principles, and techniques of value management as they relate to projects. In particular, it describes key features of the value process and the application of VM to that process, with an emphasis on the importance of value planning, teamwork, and perseverance. In addition, the incentives and benefits to all stakeholders are identified and discussed.

8.1 CONCEPT OF VALUE

The concept of value can be defined as the relationship between satisfying an organization's many conflicting needs and the resources required

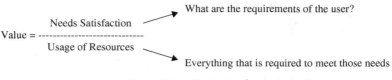

Figure 8.1 Project value[1]

to meet those needs. The fewer the amount of resources used or the higher the satisfaction, the greater the value. At the same time, many different stakeholders, including internal and external customers, may have different perceptions of what constitutes value. The goal of value management is to balance these differences and enable an organization to achieve maximum progress toward its stated goals with the minimum use of resources (see Figure 8.1).

Value can be added to projects in several ways. These include providing greater levels of client satisfaction, maintaining acceptable levels of satisfaction while lowering resource expenditures, or some combination of the two. It is also possible to improve value by simultaneously increasing satisfaction and resources, provided that satisfaction increases more than the resources used to achieve it.

When managing projects for value, five fundamental concepts must be embraced. These concepts are illustrated in Figure 8.2.

Concept #1: Projects derive their value from the benefits the organization accrues by achieving its stated goals—Remember that projects are typically initiated as a perceived solution to a goal, need, or opportunity. Thus, when we want to determine the degree to which a project is being

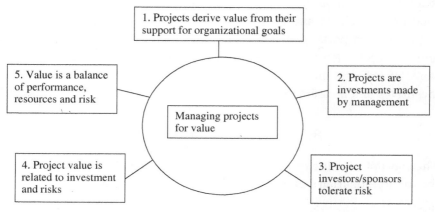

Figure 8.2 Concepts in managing projects for value[2]

managed for "value," it is critical to first ensure that the project falls in line with organizational goals. Projects that are being run counter to a firm's stated goals (e.g., customer satisfaction, commercial success, or improving health and safety) already fail the first test of value.

Concept #2: Projects can be viewed as investments made by management—Any investment comes with an expected return for the risk undertaken, and projects are no exception. Because they consume resources and time, they are expected to yield acceptable returns, based on internal requirements, along with associated benefits.

Concept #3: Project investors and sponsors tolerate risk—There are inherent risks in projects, because there is considerable uncertainty surrounding their outcomes. These risks may be technical ("Does the technology that is driving the project work?"), they may be commercial ("Will the project succeed in the marketplace?"), they may involve health and safety issues ("Can we manage the project within appropriate safety parameters?"), or some combination of all of the above. Accepting these risks is recognition that each project is a unique endeavor with unique unknowns. Investors may not be able to manage these risks, but they do tolerate them because the potential rewards may far outweigh the negative impact.

Concept #4: Project value is related to investment and risks—This fourth concept defines project value as a function of the resources committed (investment made) and the extent of risks taken. Goodpasture puts it this way: "The traditional investment equation of 'total return equals principal plus gain' is transformed into the project equation of 'project value is delivered from resources committed and risks taken.'"[3] Using these terms, we can see that value will always walk a narrow line between expected return on investment and risk. When the equation gets out of balance; that is, when the perceptions of the organization are that the expected return cannot make up for excessive levels of risk, the project ceases to produce value. The implication of this concept is that different projects require different levels of investment with varying levels of risk. Consequently, the value delivered by each of these projects will also vary.

Concept #5: Value is a balance among the three key project elements: performance, resource usage, and risk—Again, if we employ a "ledger" mindset, we can add up the credit column to include drawbacks such as expenditure (resource usage) and risk accepted. Balanced against these "negatives" is the company's expectation of project performance and positive outcomes. Naturally, the higher the expected performance of the project, the greater the resource usage and risk a company is willing to commit.

It should be clear from reading Goodpasture's perspective on value that an organization is constantly reassessing value in two ways. The first is to consider one project at a time. Each brings its own promise of value, and it is up to top management to sift through the pros and cons for each opportunity when deciding on a project investment strategy, or when forced to choose among competing project options.

The second is to reaffirm value during the project's development cycle. A project may have shown promise of delivering value early on, only to have that value brought into question at a later date. Many projects are terminated short of delivery if the perception of value becomes negative. This process can be compared to a set of balancing scales: in one bowl, we place our best guess as to a project's real benefits to the organization, and then weigh it against the risks and costs that we expect to accrue in consequence. If the scale still tips in the direction of positive outcome, the project provides positive value for the organization.

8.2 DIMENSIONS AND MEASURES OF VALUE

From Figure 8.2, it is apparent that value is a multidimensional concept that can be viewed from many different perspectives. How an organization chooses to measure value must correlate with preferred goals, available data, and meaningful metrics. For example, considerations of the quality dimension of value can include the project's conformance to standards, its functionality and fitness for use, its availability, and its responsiveness to the context and the environment. Other examples include cost effectiveness and efficiency related to project outcomes, or the benefits that the customer receives for the price paid.

For most project organizations, the objective measure of value is in monetary terms. This is because projects consume time, and the longer it takes to complete the project, the less valuable the money spent. Therefore, when evaluating a project's value, the concept of the time value of money should be taken into account. In addition, monetary measures allow project managers to acknowledge the future uncertainty of outcomes by evaluating the financial impact of risk events.

The measures of net present value (NPV), economic value add, and expected monetary value (EMV) take both of these factors into account. Net present value, which was covered in Chapter 6, is significant in that successful projects have a positive NPV over their life cycle. Both NPV and economic value add employ the discounted cash flow concept;

however, economic value add is a financial measure of project performance *after* the project becomes operational. It is defined as the difference between the present value of after-tax earnings from the project and the benefits from the next-best investment alternative. The logic underlying economic value add is that if the projected after-tax earnings are less than the cost of the capital the project consumes, then some other, less-risky investment alternative may be more attractive.

Expected monetary value is also the most appropriate financial measure when measuring future uncertainty, or when multiple project outcomes are possible, each with a different cost and schedule. It is defined as the summation of the value of each outcome in dollars ($), weighted by the probability of that outcome. For example, consider a project that needs to be redesigned, and assume that the new approach involves some risk to accomplish this goal. One possible monetary outcome is $200,000 with a 40 percent probability of achieving this outcome, while another monetary outcome is $150,000 with a 60 percent probability of occurrence. The EMV of this project is

$$EMV = \$200{,}000 * 0.4 + \$150{,}000 * 0.6 = \$170{,}000$$

If the NPV for this project is also positive, and if all other considerations are equal, it is worth taking the risk of adopting the new approach to redesign this project.

8.3 OVERVIEW OF VALUE MANAGEMENT

The concept of value and its role in the discipline of management first emerged after the Second World War. General Electric Company, in its efforts to reduce costs, undertook a comprehensive study that examined the various materials and processes that were used in producing the products required during the war. Results of the study indicated that, in most cases, the substitute materials used in production were cheaper than the original ones and performed just as well or better. This surprising finding led General Electric to develop a program that mandated that all possible substitute materials and parts should be considered for any new product development process. The study that General Electric undertook is known as value engineering and in the last 20 years its application has been extended to a number of other business processes. Value engineering is now considered to be a part of value management.

Value management (VM) is a management approach that focuses on motivating people, developing skills, and fostering synergies and innovation, with the ultimate goal of optimizing overall organizational performance. Lawrence D. Miles pioneered value management in the 1940s and 1950s, when he developed and used the technique of value analysis (VA) to enhance the value in existing products.

8.3.1 Definition

Because there are no universally accepted definitions of VM in projects, a number of different ones have been proposed to describe the same approach or stage of application. The Institution of Civil Engineers (ICE) offer a useful definition in their guide: "Value management addresses the value process during the concept, definition, implementation and operation phases of a project. It encompasses a set of systematic and logical procedures and techniques to enhance project value through the life of the facility." While the initial thrust of VM was identification and elimination of unnecessary costs, it was later applied to include services and projects.

8.3.2 Scope

Value management comprises value studies of organizational structures, management systems, and other similar activities. It is used by electronics, general engineering, aerospace, automotive, construction, and increasingly by service industries. VM techniques have been successfully applied on all types of construction, from buildings to offshore oil and gas platforms, and for all types of clients, from private industry to governmental organizations.

8.3.3 Key Principles of VM

The value management approach hinges on three key principles[4]:

1. An unending quest for enhancing value for the organization, establishing metrics or estimates of value, and monitoring and controlling them
2. A focus on clear definition of objectives and identification of targets before seeking solutions
3. A focus on function, pivotal to maximizing those outcomes that are innovative, meaningful, and practical

8.3.4 Key Attributes of VM

Value management is an important part of the overall project management mindset in that it requires us to simultaneously consider perspectives that are technical, strategic, and behavioral. These attributes include the following[5]:

1. *Management style*
    - Emphasis on teamwork and communication
    - An atmosphere that encourages creativity and innovation
    - Focus on customer requirements
    - Evaluate options qualitatively to enable robust comparisons of options
2. *Positive human dynamics*
    - Encourage people to work together toward a common solution
    - Bring people together by improving communication between them
    - Foster better common understanding and provide better group decision support
    - Challenge the status quo to bring about beneficial change
    - Assume ownership of the outcomes of value management activities by those responsible for implementing them
3. *Consideration of external and internal environment*
    - Take into account preexisting conditions external to the organization over which managers may have little influence
    - Consider internal conditions that managers may or may not be able to influence
    - Understand degrees of freedom—the external and internal conditions will dictate the limits of potential outcomes and should be quantified

8.4 VALUE MANAGEMENT TERMS

Value management refers to the full range of value techniques available, several of which are described in this section. In the past two decades, there has been an increasing trend toward applying these techniques at the earlier stages of a project's life cycle.

Value planning (VP) is a value study that occurs during the early design or development stages of a project life cycle, before a preferred alternative is selected. Value planning typically focuses on identifying project objectives and developing functional components and general approaches to meeting those objectives. It ensures that value is planned into the project from its inception by addressing and ranking stakeholders' requirements in order of importance. This makes it extremely important for project team members to know who those stakeholders are. Value planning should be used for most projects.

Value engineering (VE) is the title given to value techniques applied during the design or engineering phases of a project. This value study is conducted after the design alternatives have been developed, and perhaps before a preferred alternative has been selected. Because more information becomes available about the project as the project design process progresses, VE studies are much more detailed than VP studies.[6] Value engineering employs many techniques that focus on quantifying and comparing—it investigates, analyzes, compares, and selects among various options that will meet the value requirements of stakeholders.

Value engineering is also a useful technique for assessing costs and benefits during each phase of a project. It can be used to determine if there is a more cost-effective way to obtain the desired result, which can have a significant impact during a project's design phase.

Value analysis (VA) refers to value techniques that are applied retrospectively. Value analysis analyzes or audits a project's performance by comparing a completed, or nearly completed, design or project against predetermined objectives. Value analysis studies are normally conducted during the post-manufacture/construction period, when a project is fully operational. In addition, the term "VA" can be applied to the analysis of nonmanufacture/construction-related procedures and processes, such as studies of organizational structure, or procurement procedures.

The major emphasis of value planning, value analysis, and value engineering methodologies is to focus on value by improving existing products or developing better ones. The difference is that VA techniques focus on improvement, while VE applies value techniques for new product development.[7] In addition, VP and VE are applied mainly in the conceptual and definition phases, and generally end when the design is complete and the construction/manufacturing phase has started. However, VE can also be effectively applied during manufacturing to address problems or to exploit opportunities that may arise due to feedback from the job site regarding specific conditions, performance, and methods used.

8.5 NEED FOR VALUE MANAGEMENT IN PROJECTS

There are a number of reasons to use VM. The first is that projects often suffer from poor definition due to inadequate time and thought given at the earliest stages. The effect is seen in cost and time overruns, claims, long-term user dissatisfaction, and excessive operating costs. Value management combats this by forcing organizations to identify the need for, and scope of, any project at the earliest possible stage.

Second, there are almost always elements that contribute to poor project value, including inadequate time availability, inertia, lack of or poor communications, attitudes and influences of stakeholders, outdated standards or specifications, lack of work coordination, absence of state-of-the-art technology, and so on. Value management and its associated methodologies consciously attempt to eliminate these problems by developing an understanding of the VM processes in all project participants.

Third, VM aims to maximize project value within time and cost constraints, without any adverse impact on performance, reliability, or quality. Fourth, VM is structured, auditable, and accountable. Finally, it is multidisciplinary, seeking to maximize the creative potential of all departmental and project participants working together.[8]

8.6 THE VALUE MANAGEMENT APPROACH

Earlier, we noted that VM is based on three root principles that focus on continuous improvement of value, stakeholders' needs, and critical project success factors. These principles all emphasize the purpose of the product, system or service. Underlying these principles are three key concepts: cross-functional framework, the use of functions, and a structured decision process.[9] When these concepts are used together, they foster learning, enhance innovation, and provide an undeniable source of competitive advantage.

8.6.1 Cross-functional Framework

Decisions in project organizations can be fraught with a considerable degree of uncertainty, as well as a lack of time and quality information. The VM methodology addresses this through a multidisciplinary team approach that involves exploring every angle and sharing opinions to help the team arrive at a consensus.

The group decision-making process inherent in value management's cross-functional framework improves the quality of information and facilitates consensus and buy-in. Team members chosen from key stakeholders use the scope definition document, feasibility studies, and risk management concepts to improve the quality of information available and reduce uncertainty. In addition, the participative group decision-making process enables team members to consider more resource-effective options, and greatly enhances decision support and implementation success.[10]

8.6.2 Use of Functions

Instead of using predefined solutions, VM uses functions to determine the expected outcome of a project. This approach facilitates evaluation of much broader range of alternatives, which, in turn, enhances effectiveness and competitiveness. The concept of functions, which refers to a project's expected benefits and **critical success factors (CSFs)**, is pivotal to all value management methodologies. Critical success factors are key issues that should be addressed in all projects.[11] In the context of value management, they are the expected project benefits, defined qualitatively and ranked by project stakeholders. When these expected benefits are defined in quantitative and measurable terms, they become the project's **key performance indicators (KPIs)**, including time, cost, and quality or functionality.[12]

8.6.3 Structured Decision Process

Value management uses a structured decision-making process to ensure that the entire scope of the VM study is covered in the best possible sequence. It also addresses the increasing need to make decisions in complex environments that are characterized by considerable uncertainty. The structured decision process encourages creative thinking and efficient use of time that eventually leads to better business decisions.

It enables decision makers to engage in innovative problem-solving and decision-making techniques that solve unique problems, and provides them with authority and control over the resources needed to implement these decisions.[13]

8.7 THE VM PROCESS

The detailed VM process and the typical terms used at different stages of a project are illustrated in the Figure 8.3. Because the elements of the figure are straightforward, we will focus our discussion of the process from a strategic perspective. The elements of value management can be grouped into five major categories:[14]

- *Needs assessment*—This phase is concerned with arriving at a shared understanding of the needs of various project stakeholders, the critical success factors with the expected benefits defined qualitatively, and the key performance indicators of time, cost, quality, or functionality, defined through quantitative measures.
- *Idea generation*—In this phase, the cross-functional team focuses on generating creative and innovative alternatives to complete the project.
- *Detailed evaluation*—During this phase, the alternatives generated in the previous phase are evaluated in detail in terms of their feasibility, achievability, and potential contribution to expected project benefits. At this stage, modifying alternatives to develop additional options is also considered.
- *Optimum choice*—This phase prioritizes the various alternatives and selects the best alternative.
- *Feedback and control*—This phase is the formal evaluation and control process with feedback loops to improve the overall VM process. During this phase, VM practitioners and users must obtain feedback on its performance to ascertain whether the expected improvement in value was realized, and to generate other good ideas that can be adopted and implemented. During the feedback process, factors to be assessed include the stakeholders' judgment, involvement, and support; the system's appropriateness, use, and effectiveness; and change management. Actions that can emerge from the feedback stage include changes to personnel, approach, or the system, and even repeating the VM exercise as a whole.

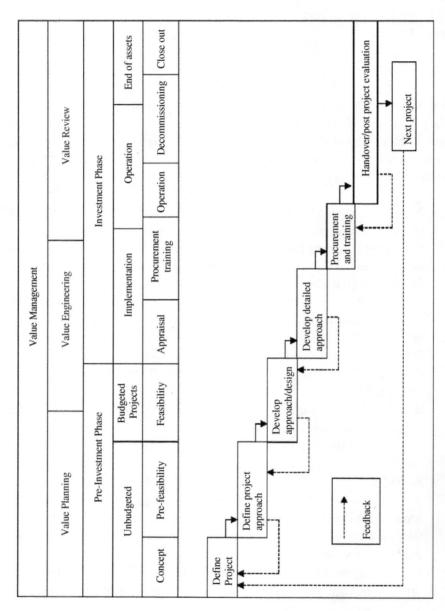

Figure 8.3 Value management process[15]

8.8 BENEFITS OF VALUE MANAGEMENT

If properly organized and executed, VM will help stakeholders achieve value for their money by striking the desired balance between cost and functional performance. This, in turn, delivers the optimum solution for project stakeholders. Value management achieves this result by ensuring that the need for a project is always verified and supported by data, and that project objectives are openly discussed and clearly identified.

The key to effective VM is to involve all appropriate stakeholders in the process of structured team thinking, so that the needs of the main parties can be accommodated wherever possible. In fact, VM primarily depends on whether or not stakeholders can agree on the project objectives from the start. Consequently, project design evolves from an agreed-upon framework of project objectives, and the outlined design proposals are carefully evaluated and selected on the basis of well-defined performance criteria.

Furthermore, during the VM process, key decisions are rational, explicit, and accountable, and alternative options are always considered. The cross-functional framework of VM ensures that there is improved communication, teamwork, and a shared understanding among the key participants. Ultimately, the cross-functional framework, the structured decision-making process, and the functional emphasis of value management ensure that project definition is of high quality, innovation is enhanced, and unnecessary costs are eliminated.[16]

8.9 OTHER VM REQUIREMENTS

There are several important additional requirements for a successful VM implementation. First, unwavering and visible support from top management is an absolute prerequisite for the success of a VM program. Second, it is imperative that all stakeholders (investors, end users, project team, the owner, constructors/manufacturers, designers, specialist suppliers, etc.) be involved in the process, especially during the VP and VE stages. The extent of their involvement depends on the level and stage of the value management process, as well as the person responsible for VM procedures.

Third, in larger or more complex projects, an independent/external value manager is needed, as well as an external team with relevant design and technical expertise. Projects that are highly complex are simply too challenging or massively complex to enable one individual to conduct all VM analyses herself. It is important to ensure that when a value manager is appointed to oversee a project, he is afforded the cooperation of all stakeholders. In the case of smaller projects, VM might be undertaken by the sponsor's professional adviser or project manager and, in some cases, an external professional. However, when establishing a structure for dealing with value for money on projects, there may be a need for expert assistance, particularly at the "review" stages.

Fourth, because each project is different, there is no single correct approach to VM. However, there are a number of stages common to all projects. Some of these stages overlap, depending on the type of project. For this reason, the project sponsor should prepare an early draft of the **project execution plan** that establishes a series of meetings and interviews, a series of reviews, a list of who should attend the meetings, and the purpose and timing of reviews. It should not be a rigid schedule, but a flexible plan, regularly reviewed and updated as the project progresses.

A typical systematic approach to developing a project execution plan begins by identifying what is to be achieved, key project requirements, characteristics, and risks. Next, data are gathered about client needs, values, costs, risks, and constraints. After this is complete, alternative options to meet the needs are generated, and risks are identified; followed by evaluation of the options defined in the speculation phase. The process ends with the identification and appraisal of the most promising options.[17]

One final requirement for value management is that the project sponsor and the value manager prepare for reviews by deciding on the objectives and outputs required, the key participants, and what will be required of them at different stages. This review process is detailed in the following section.

8.10 VALUE MANAGEMENT REVIEWS

Although the precise format and timing of reviews will vary according to particular circumstances and timetables, it is important to recognize that too many can disrupt and delay the process, especially at the feasibility stage. On the other hand, too few can result in lost opportunities for improving the definition and effectiveness of proposals.

To exploit the benefits of VM while avoiding unnecessary disruption, most projects include at least seven stages of opportunities for reviews, such as

- *The concept stage*—where reviews help identify the need for a project, as well as its key objectives and constraints
- *The prefeasibility stage*—where reviews help evaluate the broad project approach/outline design
- *The feasibility stage*—where reviews help to evaluate design proposals
- *The detailed design (appraisal) stage*—where reviews help evaluate key design decisions as design progresses
- *The manufacture/installation stage*—where reviews help to reduce costs or improve build/assembly/fabrication and functionality
- *The commissioning/operation stage*—where reviews can improve possible malfunctions or deficiencies
- *The decommissioning/end of assets stage*—where reviews can help teach lessons for future projects

All VM reviews should be structured to follow the project plan, and all elements of the project, including issues of assembly, safety, quality, operation, and maintenance should be considered during the review. Table 8.1 shows the various aspects of all seven review stages, each of which has a clear objective. For example, at the end of the first review, there should be a balanced statement of need, objectives and priorities, agreed upon by all stakeholders, which helps the project sponsor produce the project brief. Similarly, during the third phase/feasibility stage, between 10 and 30 percent of the design work is usually completed. That is why VE techniques are also part of this stage.

Value engineering is primarily aimed at finding the engineering/construction and technical solution to help translate the value planning-selected scheme design into a detailed design that provides best value. It also focuses on analyzing, evaluating, and recommending construction/assembly/fabrication proposals so that problems that may emerge during manufacture/construction can be addressed. By reviewing design proposals in this way, the value team can determine the project's purpose, functionality, cost and value, and alternatives to the project and their cost.

Table 8.1 Value management review stages

First review	1. List all objectives identified by stakeholders.
	2. Establish an objectives hierarchy by ranking the objectives in order of priority. It is important to emphasize at this point that the aim is to produce a priority listing, not simply to drop lesser priorities. Reducing the list runs the risk of having to reintroduce priorities at a later stage with all the associated detrimental impacts on cost, time, and quality. VM aims to eliminate the need for late changes.
	3. Identify broad approaches to achieving objectives by brain-storming.
	4. Appraise the feasibility of options: reject/abandon, delay/ postpone.
	5. Identify potentially valuable options.
	6. Consider and preferably recommend the most promising option for further development.
	The end result of the first review should be
	a. Confirmation that a project is needed.
	b. Description of the project, i.e., what has to be done to satisfy the objectives and priorities.
	c. Statement of the primary objective.
	d. Hierarchy of project priorities.
	e. Favored option(s) for further development.
	f. Decision to proceed.
	g. Decision to reject/abandon or postpone/delay, if necessary.
Second review	1. Review the validity of the objectives hierarchy with stakeholders and agree modifications.
	2. Evaluate the feasibility of options identified.
	3. Examine the most promising option to see if it can be improved further.
	4. Develop an agreed-upon recommendation about the most valuable option that can form the basis of an agreed-upon project brief.
	5. Produce a program for developing the project.
	The second review should result in
	a. A clear statement of the processes to be provided and/or accommodated.
	b. A preferred outline of the design proposal.
	c. The basis for the continuation of design development.
Third review	1. Review project requirements and the hierarchy of objectives that were agreed upon at the last review.
	2. Check that the key design decisions taken since the last review are relevant to the objectives hierarchy and priorities.
	3. Review key decisions against the project brief by brainstorming in order to identify ways of improving design proposals outlined to date and to identify options.
	4. Reevaluate options to identify the most valuable one.

Table 8.1 Continued

	5. Develop the most valuable option to enhance value by focusing on and resolving any perceived problems.
	6. Agree to a statement of the option to be taken forward and to a plan for the continued development of the design.
	The third review should result in
	  a. A thorough evaluation of the sketch design.
	  b. Clear recommendations for the finalization of the sketch design.
	  c. The basis of a submission for final approval to implement, abandon, or postpone the project.

Fourth review	1. Promote a continuous VM approach throughout the design process. 2. Finalize the proposed design and any changes based on the findings of the previous review. 3. Describe the value proposals and explain the advantages and disadvantages of each in terms of estimated savings, capital, operating and life-cycle costs and improvements in reliability, maintenance, or operation. 4. Predict the potential costs and savings and the redesign fee and time associated with the recommended changes. 5. Lay out the timetable for owner decisions, implementation costs, procedures, and any problems (such as delays) that may reduce benefits.
Fifth review	1. Promote a continuous VM approach throughout the construction process. 2. Assess and evaluate the contractors' change proposals. 3. Investigate and verify the feasibility of significant changes and the cost savings claimed, as well as the program implications of including them. 4. Make proactive, practical recommendations for improvements that can be implemented immediately. 5. Ensure that the risks to the project are being adequately managed.
Sixth review	1. Measure the success of the project in achieving its planned objectives. 2. Identify the reasons for any problems that have arisen. 3. Determine what remedial actions should be taken. 4. Consider whether the objectives of the users/customers have been met, and those objectives have changed or were anticipated to change during the course of the project, assess whether those changes were accommodated. 5. Ensure that any outstanding work, including defects, has been remedied. 6. Record the lessons that have been learned to improve performance on subsequent or continuing projects.

8.11 RELATIONSHIP BETWEEN PROJECT VALUE AND RISK

A project's net return and risks are the two pivotal factors that determine its ultimate success and the value it delivers. In the absence of thorough risk assessment and proactive risk management at the valuation and implementation stages, project success cannot be achieved.[18]

A simple example will serve to illustrate the importance of risk assessment and risk management to project value. Let us consider a project proposal with an expected gross return of $200,000 over five years, with a cost of $150,000 to implement. Without factoring any risks into the equation, the net return is $50,000 ($200,000 − $150,000). However, risks to a project are inevitable, and this hypothetical project is no exception. Therefore, let us assume the following risks and their impact on this project:

1. There exists some uncertainty in the project requirements and there is 40 percent probability that development efforts will cost an additional $30,000. This will reduce the net return by $12,000 ($30,000 ∗ 0.4).
2. The project team believes that there is 20 percent likelihood that additional sales force training may be required. This will likely reduce the net return by $10,000 ($50,000 ∗ 0.2).
3. There is a 10 percent probability that the entire project could fail or be superseded by other projects because of technological uncertainties or a strategic change in direction. This implies a net reduction of $5,000 ($50,000 ∗ 0.1) in the project's expected net return.

When the impact of all of the above risks is factored into account, the reduction in the net return to the project is $27,000 ($12,000 + $10,000 + $5,000), and the overall net return from the project now is $23,000 ($50,000 − $27,000). The project is now considerably less attractive than it originally appeared.

Next, we'll look at how managing the above risks will increase the value of the project to a level that makes it attractive again.

1. Let's assume that the requirements can be tightened, either by developing a proof-of-concept or by simply delaying the project until the uncertainties surrounding requirements are eliminated. By using this approach, the value of the project is increased by $12,000.[19]

2. Further evaluation of the proof-of-concept demonstrates that the anticipated additional sales force training is no longer needed. This will eliminate the second risk and increase project value by $10,000.
3. While the third risk associated with external uncertainties cannot be eliminated, it is possible to mitigate the impact of these risks on overall project value. This can be accomplished by implementing parts or stages of the project. If we assume that a fourth of the work can be salvaged, the overall project value increases by $1,250 ($5000 * 0.25).

As a consequence of risk management, the total increase in project value is now $23,250 ($12,000 + $10,000 + $1,250), and the total value of the project after risk management is $46,250 ($23,000 return after risk assessment + $23,250 total increase in return after risk management).

The example illustrates that merely acknowledging risks and building them into the cost of the project will always reduce value and lead to unrealistically optimistic project outcomes. If risks are managed proactively by formulating well-thought-out mitigation and contingency plans, their negative impact on project success can be minimized. The ultimate result is that the project team gets a more accurate and consistent assessment, and there is an overall increase in project value.[20]

8.12 VALUE MANAGEMENT AS AN AID TO RISK ASSESSMENT*

To ensure that potential risks and their consequences are identified and assessed, a qualitative risk assessment should be prepared and utilized at each VM review. A typical qualitative risk assessment usually involves consideration of the following issues: a brief description of the risk, the stages of the project when the risk may occur, the elements of the project that could be affected, the factors that cause the risk to occur, the relationship with other risks, the likelihood of risk occurring, and how the risk could affect the project.

A key factor when making decisions about risk is probability. Possible consequences are defined and quantified in terms of project cost overruns, delay in the completion date, and reduction in quality and performance. (The above impacts of risk can be analyzed using sensitivity and probability analysis.) Having identified and analyzed these risks, the

---

*This section is adapted from "Cost and Value Management," UMIST Module, Section 4.6.

value management team can then respond to them. These responses will normally entail additional costs associated with risk reduction, transferring the risk, and any retained element. In addition, as a consequence of the risk, there could be potential losses or claims based on contractual agreements. Typically risk response strategies include

- *Risk avoidance*—where risks have serious consequences on the project's outcome. Risk avoidance can include replacing a project, procuring the project under a different form of contract, or cancellation.
- *Risk reduction*—which can take the form of more detailed design or further investigation, different materials or technologies, different methods of construction, changing the project execution plan, or changing the contract strategy.
- *Risk transfer*—which can be from client to contractor, contractor to subcontractor(s), or through insurance contracts.

8.13 AN EXAMPLE OF HOW VM AND RISK MANAGEMENT INTERRELATE

Risk identification plays an important part at each review stage of the VM process, and response to risks at each review is an integral part of achieving project value. This section highlights the steps involved in effectively coordinating VM and risk management in a manner that allows them to work in synergy. The steps are highlighted in terms of the reviews found in Table 8.1.

During the first review, all potential risks are identified. While an assessment of all the risks should be done in the form of cost, performance and time, and contingency allowances, the emphasis should be on the identification of major risks. The stakeholders should be aware of the impact these risks may have on the project, and have alternative strategies for project implementation. A **risk register** should now be prepared.

At the second review stage, consideration of the various options faced by the project team (to delay, defer, abandon, or reject) includes the combined results of the risk assessment studies, the review of the reliability of data and assumptions involved, and the areas of uncertainty to be taken into account. These studies can be of significant benefit, and can help the value team form a view on possible changes to the project concept. The purpose of applying risk assessment at this stage is to provide stakeholders with a clear insight as to the major sources of risk(s) that must be taken into account; the decisions that must be made between alternative project schemes; whether or not the project can be economically justified; and the level of financing required.

During the third review, the definition of design, estimates, and program based on increasingly reliable and refined data warrants the use of quantitative risk analysis techniques. For major projects, probabilistic analyses can be conducted in areas of uncertainty. These analyses can be of significant benefit, and can help the value team to form a view on possible changes to the project concept.

During this stage, decisions should be made as to how the detailed design and implementation work should be packaged, together with responsibilities and allocation of risks to each party involved in the project. Evaluations should be made of the many elements that remain undefined at the end of the conceptual phase, and which relate to all subsequent work up to the final commissioning. In addition, the effects of unforeseen risks that arose between the previous and current review should be quantified, and appropriate risk responses should be developed and implemented.

Again, the various options faced by the value team are based on the combined results of the risk assessment studies, the review of the reliability of data and assumptions involved, and the areas of uncertainty that need to be taken into account. At the conclusion, a review and evaluation of the plan should be made, followed by recommendations for further adjustment, improvement, or changes.

In the end, project value is based on three important aspects: the independence of the value manager to clearly establish the stakeholders' value criteria, planned application of team brainstorming, and the inclusion of appropriate enabling clauses in contracts and agreements. The factors needed to ensure value management success include a systematic approach, an integrated team environment, establishment of value criteria, focusing on the function, consideration of project life-cycle costs, a collaborative and nonconfrontational working environment, generation of records, and an audit trail. Value management must also have comprehensive understanding and support at the top levels of management, as well as an enthusiastic, sustained, and innovative approach to implementation. When properly organized and executed, VM provides a structured basis for both appraisal and development of a project, and results in many benefits.

The process of developing an integrated VM system is not simple, and it does not allow for quick fixes or easy definition of risks and value. The key point to remember is that value represents an organization's commitment to achieving the best possible return for its projects relative to expenditures, risks, and other costs.

REFERENCES

1. (n.d.) http://www.ivm.org.uk/vm_whatis.htm
2. Goodpasture, J. C. (2002) *Managing Projects for Value.* Vienna, VA: Management Concepts Inc., p. 2.
3. Goodpasture, J. C. (2002) *Ibid.,* p. 4.
4. (n.d.) What is value management. The Institute of Value Management. http://www.ivm.org.uk/vm_whatis.htm
5. (n.d.) What is value management. The Institute of Value Management. http://www.ivm.org.uk/vm_whatis.htm
6. (n.d.) Reclamation value program's frequently asked questions. Bureau of Reclamation, Department of the Interior. http://www.usbr.gov/pmts/valuprog/faq.html
7. Thiry, M. (2004) Value management. In: P.W.G. Morris, and J. K. Pinto, (Eds.). *The Wiley Guide for Managing Projects.* Hoboken, NJ: Wiley, pp. 876–902.
8. Cost and value management. UMIST Module, p. 4.6.
9. Thiry, M. (2004) *Ibid.*
10. Thiry, M. (2004) *Ibid.*
11. Pinto, J. K., and Rouhiainen, P. J. (2001) *Building Customer-based Project Organizations.* New York: Wiley.
12. Thiry, M. (2004) *Ibid.*
13. Thiry, M. (2004) *Ibid.*
14. Thiry, M. (2004) *Ibid.*
15. Cost and value management. UMIST Module, p. 4.6.3.
16. Cost and value management. UMIST Module, p. 4.6.
17. Cost and value management. UMIST Module, p. 4.6.
18. Dandrea, R. (n.d.) Increasing project value through risk management. http://www.processimpact.com/articles/risk_mgmt.html
19. Dandrea, R. (n.d.) *Ibid.*
20. Dandrea, R. (n.d.) *Ibid.*

KEY TERMS

Value
Value management (VM)
Project value
Value planning (VP)
Value engineering (VE)
Value analysis (VA)
Critical success factors (CSFs)
Key performance indicators (KPIs)
Project execution plan
Risk register

Chapter 9

Change Control and Configuration Management

LEARNING OBJECTIVES

- Discuss the project control system called configuration management.
- Analyze the four stages of configuration management.
- Describe change control procedures.

One of the inevitable features of project management is change, defined as any addition, deletion, or alteration to the scope, nature, quantity, standards, timing, or location of work for a project during implementation. Changes can be frustrating to project managers and teams because they are usually the cause of significant delays and extra costs, particularly when they lead to contract variations.

Even so, not all changes are, of themselves, a bad thing. Changes can provide opportunities to improve a project, or to overcome problems. They can occur as the result of new technologies or innovative solutions that will result in a stronger end product. In fact, changes reflect the fact that projects are developed in a dynamic environment that is constantly offering new opportunities and threats. If we objectively consider the implications of change—instead of responding with the typical knee-jerk reaction of frustration when project change requests are made—there are many strong reasons why project changes are a positive for the organization.

In this chapter, we discuss how to manage and control changes so that project objectives can be met. We talk about why changes occur, and explore their implications on project cost and value. We discuss crisis management, and whether changes to a project can prevent it from achieving its defined objectives after the project is initiated. We also explore two important processes: the project change control system and, particularly, a component of this system known as "configuration management" (CM) that enables the compilation, control, and communication of critical project information.

9.1 CAUSES OF CHANGES

Recently, a simple study was conducted to assess the causes of changes during the implementation of projects. Data were obtained from responses to the following two questions[1]:

1. What is the greatest problem you face in project management?
2. In your opinion, what is the cause of that problem?

The results, from the perspectives of people from a number of different industries, are presented in Table 9.1.

The study also found that the causes of changes listed in the table above were primarily due to

- Markets and political conditions, such as exchange rates, material supplies, and other input factors.
- Project implementation conditions and resources that differ from the best predictions.
- Project sponsors who tend to spend as little time as possible on investigations before a project is authorized. (This is because cost and other resources appear to be wasted if the project does not go ahead, or if it is changed by any of the above factors. As a result, many problems are perceived and options considered only after the decision is made to go ahead, and therefore are incurred at the expense of project progress.)
- Project sponsors who are tempted to make changes to incorporate new technology, although this was not the original project objective.
- Contract and internal procedures that evolve to manage essential changes and variations, which create an expectation that specifications and other decisions are also open to change.

As mentioned earlier, change can be an advantage when it paves the way to using new ideas, and to developing and applying innovations during a project's implementation stage. Changes can be classified as

- *Changes inherent in projects*—These include changes due to markets, customer demands, evolving relationships with customers and tasks from other projects, immature technology, new information, and the nature of development work.

Table 9.1 Causes of changes during project implementation

Poor communications
Culture and structure of organization, personalities
Pressure/time constraints preventing effective definitive design
Changing markets, customer demands, ideas
Company attitude, poor discipline
Innovative, immature technology
Contracts signed when scope is not frozen
Schedule-driven projects
Customers changing their minds
Conflicting requirements within the team, lack of experience
Getting contractors to "buy into" plan
Insufficient front-end loading of the project
Changes from initial design
Evolving relationship with customers and tasks from other projects
Lack of preparation, inadequate risk assessment
Lack of personal knowledge/experience, time pressures
Poor initial planning
Not enough discussion among all stakeholders at the start of project
Poor scope documents, scope changes, lack of resources, time
Not enough time at the front end to finalize design
Projects are nebulous, company run by jargon-spouting accountants
Scope of work unclear at start of project, changes in market, technology
Lack of understanding by project sponsors (operations)
Developing situation in areas of new technology
Changes in design after commencing installation
Inadequate planning for business strategy
Individuals always have their own ideas on how to carry out their work
The nature of development work
The nature of the work itself
Unforeseen problems, difficulty of scoping
Being unable to see all the issues at the beginning
Lack of project management tools and understanding of workloads
Lack of understanding of the project on the part of senior management
---

- *Changes internal to the business*—These include changes due to poor communication, culture, and structure; personalities; company attitude; poor discipline; pressure and time constraints preventing effective definitive design; conflicting requirements within the team; lack of experience; poor initial planning; and most of the rest of the list in Table 9.1.

With regard to changes inherent in a project, the project sponsor must accept the changes, and the project manager must anticipate their effects. In mature project organizations, changes internal to the business are managed by reviewing previous projects that have shown what lack of action has cost.

Based on the above discussion, the need to make significant project changes comes about for one of several reasons[2]:

- *As the result of initial planning errors*—Because many projects involve significant technology risks and uncertainty, it is often impossible to accurately account for all potential problems or technological roadblocks. As a result, many projects require midcourse changes to specifications when they encounter unsolvable problems or unexpected difficulties. Planning errors may also be due to simple carelessness or a lack of full knowledge of the development process. In this case, where the causes for change are nontechnical, reconfiguration represents a simple adjustment to original plans to accommodate new project realities.
- *As a result of additional knowledge of project conditions*—The project team or client may enter into a project, only to discover that specific project features or the development environment itself require midcourse changes to scope. For example, the technical design of a deep-water oil drilling rig may have to be significantly modified upon discovery of the nature of water currents or storm characteristics, underwater terrain formations, or other unanticipated environmental features.
- *Uncontrollable mandates*—In some circumstances, events occur outside the control of the project manager or team that must be factored into the project as it moves forward. For example, the new passenger safety requirements established by the European Union temporarily forced Boeing Corporation to redesign exit features on their new 777 aircraft, delaying the project's introduction and sale to foreign airlines.
- *Client requests*—As a project's clients learn more about the project, they often ask for significant alternations to address new needs. For example, in software development, potential users usually list

complaints and requests for new features, reworked features, etc. when they are first exposed to a planned software improvement. In fact, IT projects often run excessively long because users continue to bring forward lists of new requirements or change requests as software development progresses.

Case Study: The Bradley Fighting Vehicle[3]
A famous example of the problems that can beset a project as the result of excessive change orders, evolving needs, and conflicting information is found in the development of the U.S. Army's Bradley Fighting Vehicle. Now a fixture of army inventory, the Bradley has been in service since 1981. It has been used in combat in the 1991 Gulf War, in Somalia in 1998, in Bosnia in 1999, and in Afghanistan in 2001, and in Iraq throughout the current conflict. Although it has been deployed reasonably successfully in combat, the case of the Bradley Fighting Vehicle is an example of a project whose original scope was altered so greatly that the project finally achieved a life of its own, seemingly divorced from its original goals.

Manufactured by FMC Corporation, the Bradley was originally conceived and designed in the early 1960s to replace the army's older M-113 armored personnel carrier (APC). A large, tracked vehicle with minimal offensive firepower, the APC's job is to bring troops riding inside into battle areas as quickly as possible. Once in combat, the APC works with other armored vehicles to develop and exploit breakthroughs on the battlefield.

Original specifications of the Bradley's design included the capability to transport an entire squad of infantry (12 soldiers); a top speed sufficient to keep pace with other armored vehicles; strong side armor to protect the crew and infantry riding in the vehicle; the ability to travel through water (amphibious capability); and minimal offensive firepower (to travel light). Although initial plans suggested a fast turn-around for the project, almost two decades later, the army received a vehicle that

- Could transport only six personnel
- Was so lightly armored that enemy weaponry could easily penetrate its sides
- Sank like a stone when attempting to cross rivers
- Carried a hefty, full complement of machine guns, a 25-mm cannon, and anti-tank missiles

In other words, its realization was completely out of proportion to its originally conceived role on the battlefield. What went wrong?

Perhaps the biggest problem that hampered the development of the Bradley almost from the beginning was the failure within the army to come to a definitive agreement on the role the vehicle was to play on the modern battlefield. As mentioned previously, the army began simply with the need to replace its older APC; army brass originally designed it to be deployed against Warsaw Pact forces in Europe. However, they quickly added another requirement: that the Bradley serve in the dual role of armored scout.

These two missions were not compatible. Heavy armor, large size, and weaponry only for defensive purposes were the qualities that defined an APC. An armored scout, on the other hand, needed to trade armor plating for extra speed. Further, a scout's mission was also more offensive, requiring that the vehicle be outfitted with a full complement of weapons, including a turret with cannon, machine gun, and anti-tank missiles. The army's lack of clarity for the Bradley's basic mission doomed it to serve poorly in both roles, and dramatically lengthened the development cycle. Redesign followed redesign and modification followed modification, rapidly consuming the budget and stretching development out through the administrations of five U.S. presidents.

Once put into production at FMC Corporation, the Bradley was also plagued by poor quality control. Employee whistleblower Henry Boisvert and others routinely witnessed sham testing, falsified documents, and poor quality control on the assembly lines. Army tests were conducted no more honestly, as the Air Force officer in charge of operational testing and evaluation, Colonel Burton, discovered. Some live fire tests were rigged, and some results were fabricated. To maintain an acceptable top speed, the Bradley sacrificed standard armor plating in favor of a form of aluminum that burned easily, while giving off deadly fumes. Further, to support the weapon complement necessary for the vehicle to serve as an armored scout, the interior was much smaller.

Since 1981, when the Bradley originally rolled off the assembly line, over 6700 have been put into service by the U.S. Army. From the outset, the vehicle has faced a mixed reception. Advocates argue that it serves a useful role in the modern, highly mobile army. Critics point out that as a result of its poor initial project scope and the army's willingness to make numerous, dramatic changes to the role of the Bradley midway through development, the Army was left with a dangerous lemon. The Bradley suffered from the problem of scope creep, or the continual reassessment and change of a project's original specifications. For the army, the final price tag for constant change was in excess of $14 billion dollars.

9.2 INFLUENCE OF CHANGES

Research has shown that we tend to underestimate the impact of project changes on cost and schedule for the following reasons:

- Changes are often proposed as the solution to a problem without a thorough investigation or full understanding of the difference in cost between solving the problem and living with it.

Configuration Management

- Only the direct cost of a change may be obvious. The other costs associated with the disruption of work and of relationships among project stakeholders caused by the change are difficult to estimate, and tend to become apparent only much later. Their cost impact can be much greater than those of direct costs; for example, consider situations in which costs are estimated using learning curves, as discussed in Chapter 3. Changes to the project in midstream are very likely to disrupt the positive effect of learning curves and add significantly to the project's bottom-line cost.
- Similarly, the ultimate price of ordering contract *variations* tends to be underestimated.
- People tend to be overly optimistic about the success of adopting new ideas, including changes to projects, without considering their impact on cost and schedule. However, experienced project managers are of the opinion that the potential advantages of making changes are nearly always overestimated, and do not really serve the original objectives.

9.3 CONFIGURATION MANAGEMENT

A large volume of information is generated during the implementation of a project, and both cost and value are affected by how well this information is compiled and communicated. This is particularly true of large projects that are affected by frequent changes—the larger the project, the more important it is that there is a system for compiling, communicating, and controlling the subsequent changes to information. For commercial or other security-related reasons, the system may have to control access to this information.

An additional challenge to managing project information is that it becomes much more detailed and diffused as various people with different expertise undertake their specialist activities. Clearly, everyone involved should base their work on consistent data about the project's form, size, and intended performance.

These challenges are addressed by the project change control system, and particularly by a component of this system known as **configuration management** (CM). The Project Management Institute's Body of Knowledge (PMBoK) defines configuration management as "a system of procedures that monitors emerging project scope against the scope baseline. It requires documentation and management approval on any change to the baseline."[4] (The scope baseline is simply a summary description of the project's original content and end product, including budget and time constraint data.)

In simple terms, configuration management is best understood as the **systematic management and control of project change**. It is important to recognize that the best time to start planning for change is at the beginning of the project, when scope, budgets, schedules, and activity sequences are established. On the surface, it may seem odd that the management of project changes is most effectively accomplished at the project's beginning. Nevertheless, because the need to make significant project changes is usually an acknowledged part of the planning process, configuration management seeks to formalize the change process as much as possible, rather than leaving changes needed downstream to be done in an ad hoc or uncoordinated manner.

9.4 CONFIGURATION MANAGEMENT STANDARDS

The configuration management process, as we understand it today, became its own discipline sometime in the late 1960s. In the 1970s, the U.S. Government developed and issued a series of military standards for configuration management that was consolidated into a single standard known as MIL-STD-973 in 1991. This standard is still active but due for cancellation. In the early 1990s, an initiative began to adopt nongovernment standards wherever possible; this was the beginning of what has now evolved into the most widely distributed and accepted standard on CM. Known as the "National Consensus Standard for Configuration Management," ANSI/EIA-649-1998 was developed by the Electronic Industries Alliance, reviewed by a multiassociation advisory group, and approved as an American National Standards Institute (ANSI) document. Without mandating requirements, this standard provides an excellent set of processes, principles, and guidelines for conducting CM in any environment.[5] ANSI/EIA-649 states that "the primary objective of CM is to assure that a product performs as intended and its physical configuration is adequately identified and documented to a level of detail that is sufficient enough to repeatedly produce the product and meet the anticipated needs for operation, maintenance, repair and replacement."[6]

9.5 THE CM PROCESS

It is important to understand CM as a component of the normal project development processes within an organization. Configuration management does not occupy a separate position; rather, its principles are embedded routinely within normal development cycles. According to the University of Manchester's Centre for Research on the Management of Projects (CRMP), "The goal of the CM process is to address the composition of a product, configuration documentation defining a product, and other data and products that support it. To do so, CM is structured into four integrated processes that provide for complete control of information for project life cycle management"[7] (Figure 9.1).

The specific tasks of the configuration management discipline are as follows[9]:

1. *Configuration identification*—This process identifies all items uniquely within the configuration, which establishes a successful method for requesting a change and ensures that no change takes place without authority. In choosing the names of items, it is advisable to keep them short, but unique and meaningful. In addition, every configuration item should be physically labeled so that the label identifies that physical item as the one recorded in the configuration register. Furthermore, when a change is required on a piece of fielded item, there must be a simple way, such as etching the item with numbers, to record each change made to that item. While etching numbers is easy for tangible physical items, it is not easy for intangible items like

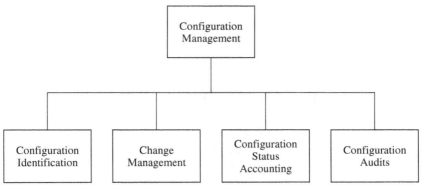

Figure 9.1   Configuration management structure[8]

software packages. Even so, there should be an identification mechanism to determine which version of the software is actually installed. In the case of large projects that involve numerous configuration items, it is important to establish a **baseline configuration** to provide some structure and avoid confusion. According to Field and Keller, "A baseline is the state of configuration, or a defined part of the configuration, at a particular point in the development, usually at a major milestone in the system life cycle. A baseline may be defined as a set of known and agreed configuration items under change control at a discrete time from which further progress can be charted."[10]

2. *Configuration control*—This is a system through which changes may be made to configuration items. As change requests begin to appear, the configuration control system ensures that no change is made without assessment of its impact, either by the people potentially affected by the change, or without approval by an appropriate authority. The element of configuration control is examined in much greater detail later in this chapter when we discuss control of changes to projects.

3. *Configuration status accounting*—This process, which records and reports the current status and history of all changes to the configuration system, provides a complete record of what happened to the configuration system to date. The purpose is to provide a mechanism to trace all of the events that happened to the CM system so that comparisons can be made with the original development plan. Again, according to Field and Keller,

> "The records that are accumulated in the configuration status accounting database include:
>
> - The creation of new configuration items together with the name of the change control authority
> - Incident reports (i.e., faults reported that may produce change requests)
> - Change requests
> - Change approval/rejection decisions
> - Change notices (issued when item have been changed)
> - Updates to items stored in the system itself
> - Baseline configurations
>
> Using these data the status accounting system should be able to produce reports on the status and history of any item or collection of items, such as a baseline."[11]

4. *Configuration audit*—These audits are performed to ensure conformity between the items in the configuration and their specifications. Audits ensure not only a match between what is delivered and what was requested, but also consistency throughout all project documents. In other words, even though changes may have taken place in the requirements and design, configuration auditing ensures that the items produced conform to current specifications, and that all quality assurance procedures that are said to be in place have in fact been performed satisfactorily. Configuration audit also ensures that changes that have been made to the items or the project are correctly recorded.

An alternative representation of the four stages in configuration management includes the steps shown in Table 9.2.[12]

Table 9.2   Four stages of configuration management

Step	Action
1. Configuration identification	1. Develop a breakdown of the project to the necessary level of definition. 2. Identify the specifications of the components of the breakdown, and of the total project.
2. Configuration reviews	Meet with all project stakeholders to agree to the current project definition.
3. Configuration control	1. If agreement is achieved, repeat the first three steps, developing the breakdown and specification further, until the project is defined. 2. If agreement is not reached, either • Cycle back to the configuration as agreed at a previous review and repeat steps 1–3 until agreement is achieved; or • Change the specification last obtained by a process change control to match what people think it should be.
4. Status accounting	Memory of the current configurations and all previous ones must be maintained, so that if agreement is not reached at some point, the team can cycle back to a previous configuration and restart from there. Also, memory of the configuration of all prototypes must be maintained.

9.6 CONTROL OF CHANGES

We have argued so far that changes to a project are inevitable. Sometimes the reasons make good business sense (e.g., significant changes to the company's strategy); at other times, they may occur for a variety of poorer reasons. Nevertheless, we have to take, as a starting point, the perspective that changes will be embedded in any project. When changes happen, the most important issue is not the degree to which they occur, but the manner in which we stay on top of them by proactively working to control the change and configuration management process. There are a number of means by which project organizations can first anticipate and then plan for and manage changes. From our own experience and research, we can highlight several suggestions for anticipating potential sources of project change:

1. Evaluate all possible options/alternatives during the stages of project scope and procurement, and during the planning phase of the project feasibility study. Apply the lessons learned from previous projects. Even under pressures to limit expenditure, this is the time to use resources to consider alternatives and preferences.
2. Foresee and list the possible reasons for potential changes. Use risk analysis to assess flexibility and margins to allow in the project size, budget, and schedule.
3. Plan the project and identify all novel and uncertain details before project sanctioning, so that the appropriate decision can be made on how to proceed in implementing the required change, or whether to go ahead with the project.
4. Involve all project stakeholders in the above decisions and obtain commitment from them, thereby motivating them to feel that they own the project definition and are personally responsible for the results.
5. If the project objective is development, proceed in stages and spend the money on work that will reduce the uncertainties associated with whether to proceed further, and how.
6. If the project is urgent, make final decisions before starting. Overlapping planning and implementation phases in a project can cause great delays.
7. If schedule or cost is important, establish a procedure to anticipate, assess, and decide on possible changes.

8. Consider whether to set a policy that no changes will be accepted unless imposed by new legislation, or unless the changes offer an overwhelming and undoubted advantage to the project, such as a significant financial rate of return.
9. Maintain continuity of the senior staff, particularly those responsible for sponsoring and managing the project.

9.7 CHANGE CONTROL PROCEDURE AND CONFIGURATION CONTROL

To avoid the costs and delays associated with uncontrolled changes, many organizations have set up formal procedures to control them—and have further extended those procedures "upstream" to anticipate changes well before they occur. Establishing a configuration control system provides just such a framework for anticipating what may not be obvious. Aided by systematic questioning and checklists from other projects, the configuration control system is designed to reveal omissions in the project definition, along with the presence of inconsistent specifications.

While many changes in projects should not be accepted, some may be required for good reasons, and implementing them may provide significant advantages. Therefore, "formal systems to control changes should be operated as an aid to improve project performance and success."[13]

The procedure to control the potential effects of changes on cost or schedule can be simple. It needs to provide a basis for making decisions in time to minimize real problems, yet deter the temptation to make a change, rather than solve an underlying problem. The procedure should require anyone proposing a change to provide the following information:

- Who wants the change, and why?
- What are its potential advantages and consequences?
- What are the potential effects on schedule and budget?
- What are the alternatives?
- Is the change being proposed to overcome a hidden problem?
- What would be the consequences of refusing the change?
- What do the other stakeholders recommend about the proposed change?
- When and how should the change best be implemented?

Change control involves organizing a collection of people (including management, the development team, testers, user representatives, and the client) to cope with change and to minimize its disruptive impact. Change control can be executed in a project environment by the following procedure:

1. Define the current state of the system to which any change is referred.
2. Establish a means for submitting proposals for change.
3. Establish a mechanism for evaluating the impact of change, including cost, time, and effects on others.
4. Have a decision-making authority to approve or reject the change.
5. Establish a method of recording the approved or rejected change.
6. Establish a method for publishing the change.
7. Establish a means of monitoring the implementation of the change.

A sample change control form is presented in Figure 9.2.

There are two other issues relating to changes in projects. The first is associated with project delays, which are a recurrent problem with many projects, and one of the critical reasons for this is significant changes made to project specifications after formal approval. This issue continues to be a major source of headache for project managers, who need to freeze specifications to minimize likely downstream project disruptions. Consequently, the control of changes needs to be both quick and in tune with overall project objectives. On the other hand, if a decision on a proposed change is delayed until the end of the project or a milestone date because it may be cheaper to implement it together with other changes, the facts could likely change by that time, and the change may become unnecessary.

The second issue relates freezing the project. It is common to hear project managers argue for freezing specifications at the earliest possible point. They note that the more "fluid" the project's specifications, the more likely that change orders will be issued, making final project completion late and expensive. On the surface, their arguments seem to have merit: specification changes are often a "killer" when it comes to managing an efficient implementation process. Further, the later into the project that these change requests arrive, the more expensive and significant their negative impact on the project is likely to be.

Change Control Procedure and Configuration Control

	TRIMAGI COMMUNICATIONS CHANGE CONTROL FORM
PROJECT: WORK PACKAGE: ACTIVITY: ORIGINATOR:	CRMO RATIONALIZATION _____ _____ _____ _____
DESCRIPTION OF CHANGE	
IMPACT OF CHANGE	
COST OF CHANGE: $ _____ VALUE OF CHANGE: $ _____	
	NAME SIGNATURE DATE
PROPOSED BY: CHECKED BY: APPROVED BY:	_____ _____ _____ _____ _____ _____ _____ _____ _____

Figure 9.2 Change control form[14]

Although this is a reasonable position to argue from, it does not ultimately hold much merit. There are several irresistible forces that can necessitate changes to the project scope, such as market conditions, actions that need to be taken to correct mistakes, and new technology. The scope and schedule for a project can be frozen only in circumstances where there are overwhelming reasons to do so, or in cases of emergency projects, where speed is the overriding basis for decisions.

The practical lesson learned from industrial and public projects is that flexibility and spare capacity are needed because of uncertainties during project implementation. However, to use them successfully, there must be a formal system that controls how and when they are used. Without that control mechanism, flexibility and spare capacity may be misused to cover up the poor decisions that were made at the project's outset.

9.8 RESPONSIBILITY FOR THE CONTROL OF CHANGES

To control changes, we need to know where they come from. To help in this effort, we can classify the causes of changes listed in Table 9.1 as originating from three different sources:

1. *Corporate management*—Examples include changes due to culture and structure, poor discipline, and immature technology. Clearly, these types of changes are a corporate rather than project responsibility.
2. *Business environment*—Examples include changes due to markets and customers' demands, and changes to take advantage of new ideas and technology. The control of these types of changes should be the responsibility of the project sponsor (whoever is responsible for initiating the project or for bidding to a potential purchaser), and allowance is usually made in the budget for such cases.
3. *Project management*—Examples include changes due to contracts signed when the project scope was not frozen, conflicting ideas within the project team, lack of preparation, and inadequate risk assessment. The control of changes due to the project environment should be the responsibility of the project manager, and is factored into the budget for project execution.

Change control in large projects is often the function of a committee. In such cases, changes in the first two categories should be decided by the committee in consultation with the project manager. With regard to changes in the third category, they should be reported to the committee for informational purposes only, and not for the purposes any decision making by the committee. The rationale is that the committee should spend their time concentrating on the more critical changes that may affect the project's objectives and priorities.

9.9 CRISIS MANAGEMENT

A project crisis is a situation where it appears that the project objectives are no longer achievable. A project crisis may cause a major change to a project; that is, one of the project's original objectives of performance, quality, schedule, or cost may have to be sacrificed. If the crisis is a threat to achieving project objectives, a decision needs to be made on whether to proceed any further with the project. When making this decision, up-to-date information on questions such as what can be done, how, when, and at what cost should be thoroughly analyzed. In essence, the solution to a crisis may become a replacement project, which should be evaluated so that a decision can be made as to whether to proceed with the current project or to abandon it.

To manage a crisis effectively, it is important that the danger signs are recognized as early as possible. Unfortunately, however, it is often true that the people working on the project may be unaware or unable to agree that a crisis is imminent. This is particularly true in bureaucratic organizations, where it may be difficult to obtain a consensus so that appropriate action can be taken in time to minimize damage. Invariably, in these types of organizations, by the time the crisis is recognized it is usually too late, and decisions on how best to respond to the crisis are usually of an urgent nature, and not well thought out. This deters the process of finding the appropriate facts, and leads to strained relationships among project team members—and the avoidable panic that has been created reduces confidence in the organization.

Ideally, to effectively manage a crisis, the project manager should be in regular and informal contact with everyone on the project so that the early warning signals of an approaching crisis are recognized. The project manager should then take a logical, step-by-step approach to crisis management by finding answers to the following questions:

1. What is the problem?
2. Why is it urgent?
3. How will the crisis affect the achievement of the project objectives?
4. Do all the concerned parties agree on the facts?
5. What are the options?
6. What is the best course of action?

Operating in crisis mode requires some clear-headed decisions and steps, the most important of which include:

- Accessing and assimilating "lessons learned" information from previous projects. One of the best defenses against crisis, and one of the best ways to establish the means to respond quickly, is to consider the crisis in relation to historical data. Have we experienced similar work conditions in the past? Does this customer or supplier have a reputation for inconsistency? Lessons learned can often serve as a guide for pointing out warning signs of potential problems, or offer clear guidance of optimal remedial steps once the crisis hits.
- Applying appropriate planning and risk analysis at the start of a project. Risk management is a key means for dodging the worst possible problems during a project's development. Obviously, risk assessment is useful only to the extent that those causes can be foreseen. Nevertheless, if we have performed a reasonable risk assessment, we should have begun developing contingency plans for many of the potential sources of project crises.
- Providing the project manager with the authority necessary to address the crisis. Many organizations do not give their project managers enough latitude to respond to project threats, preferring that all "critical" decisions be pushed up the organizational ladder. This perspective is dangerous. Due to time constraints, it may not be possible to consult with the project sponsor or other stakeholders when a crisis threatens. Therefore, the project manager needs to have or assume the authority to define the crisis and to decide on actions to be taken on behalf of the project sponsor.[15]

9.10 AN EXAMPLE OF CONFIGURATION MANAGEMENT[16]

So far, we have spoken in general terms about the development and usefulness of a formal configuration management system. Next, let's apply these principles to a simple example that will illustrate the effect of configuration management on how the project is conducted. The example—a project to develop a "green" refrigerator—and the design team involved are modest by comparison to others; however, it does illustrate principles that are equally appropriate in large and more complex projects.

An Example of Configuration Management

Example

A small but increasingly successful firm has built a good reputation for its refrigerators and freezers. The management of the firm is now convinced that there is requirement in the market for a "green," highly energy-efficient refrigerator. Since the company is small, they feel it is possible to quickly fill this expanding niche in the market with a good product. After initial analysis, a product requirement definition document is agreed upon. This becomes the first configuration item:

Configuration item

Item name:	Requirements definition
Created:	1/10/07
Change authority:	Engineering Division manager and Marketing manager
Description:	Master copy held in Engineering files ref Green/1/1
Items depended on:	None

This information is held in a database under the control of an administrator in the engineering department. The description is a pointer to show exactly which document is being referred to, and the requirement definition itself is held securely in a file. Because it is the first item, it does not depend on other items already in the system.

The design team has two lead designers who decide that one will be responsible for the refrigeration unit, the controls, and the motor, and the other for the case, the interior fittings, and the door. Together, they draw up a design and submit it to a meeting chaired by the marketing manager, where it is confirmed that this design conforms to the requirements definition. This now becomes the second configuration item:

Configuration item

Item name:	Top-level design
Created:	1/25/07
Change authority:	Lead designers with project manager
Description:	Master design held in Engineering files ref Green 1/2/1
Items depended on:	Requirements definition

Note that this item is dependent on the requirements definition, an item already in the configuration. Any request for change in the requirements

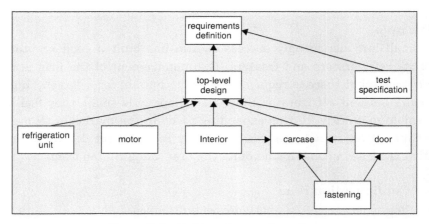

Figure 9.3 Refrigerator configuration

definition could affect this second item. As the design of individual components proceeds, other configuration items are stored in the database—the designs for the case, the motor, the interior fittings, etc. As each item is added to the database, it is necessary to record precisely where that design is held, and it must be held in such a way that no changes take place without authority. Some items spawn others (the door design, for example, has a fastening design that is dependent on the case and door designs). The configuration of the refrigerator after about a month looks like Figure 9.3.

Next, change requests start to appear. The designer responsible for the refrigeration unit undertakes some research on non-CFC refrigerants and discovers that using these coolants will require some changes. In particular, the coils have a different geometry from the one specified in the top-level design. Consequently, there needs to be a change request not only for the refrigeration unit itself, but also for the top-level design:

Change request
Originator: Abel Baker
Date: 2/15/07
Items(s) to be changes: Top-level design and refrigeration unit design
Outline of change: The refrigerator will use liquid propane as coolant, and will require a design for the refrigeration coils and a motor suitable for this coolant.
Reason for change: Liquid propane is presently the most effective non-CFC coolant available for domestic refrigeration.

An Example of Configuration Management

The administrator checks the database to find out what other items are recorded as *directly* dependent on any of the items for which the change is requested, so that the full impact of the effect of a change can be assessed. Notice that in the first instance, we consider only items that have been declared dependent on the items for which change has been requested. There could be lower-level items that depend indirectly on the top-level design, via, say, the case design, but there is no need to consider them unless a change is necessary in the case design itself.

The designers meet at the project manager's instigation and determine that the changes required by using liquid propane will affect the refrigeration unit and the top-level design, but not the case design. Nevertheless, the fact that there has been a change is reported to all members of the design team, in case someone working on the design of something that is not yet part of the configuration is relying on the earlier designs.

The administrator using the database records all change requests, impact assessments, and decisions made by item controllers. Occasionally, the designers ask the administrator questions, such as, "How did we come to use this size motor?" The administrator answers this by using the database to recall design history from the original document identified as motor design, as well as all subsequent changes to it.

The quality manager also asks questions to ensure that the product is meeting the intentions of the managers of the marketing and engineering divisions. These include

- Is there a record of all configuration items, uniquely identified, with records of all requests for change and the outcome of each?
- Have all changes been properly authorized?
- What evidence is there of verification that the detailed design conforms to the top-level design?
- Is there a record of management agreement to the top-level design? Are there any outstanding requests for change to the designs?
- Is there a specification of the tests that will be conducted before the refrigerator design is approved for manufacture?
- Overall, if the refrigerator were constructed to the detailed designs, would it meet the sponsor's requirements?

Everyone in the department who had anything to do with the development of this refrigerator was affected by the decision to use configuration management. The project manager found that he had quite a "selling" job to ensure that everyone adheres to the system, even if they don't really like it. Some found it onerous and were irritated by the extra administration and the delays caused by the change control procedures. But others found it useful that nobody changed things without letting them know, and liked the way that they could get information about the reasons for past decisions.[17]

Change control and configuration management represent critical elements of effective project management across the full life cycle. It is perhaps ironic that we should be focusing on the need to manage project changes even during the initial planning phase, but this mentality reflects the realities of modern projects—highly technical, complex, and often having multiple stakeholders. As we noted in this chapter, a variety of phenomena can have significant effects on the project, mandating the need for both minor and significant changes. With the goal of managing our projects for maximum value and cost control, configuration and change management is directly in the forefront of prudent planning and project execution.

REFERENCES

1. Cost and value management (1997) UMIST Module 4.
2. Meredith, J. R., and Mantel, S. J., Jr., (2011) *Project Management*, 8th ed., New York: Wiley.
3. Pinto, J. K. (2009) *Project Management: Achieving Competitive Advantage*. Upper Saddle River, NJ: Prentice-Hall, pp. 146–147; Burton, J. G. (1993) *The Pentagon Wars*. Annapolis, MD: U.S. Naval Institute Press.
4. Project Management Institute (2011) *Project Management Body of Knowledge*, 4th ed., Upper Darby, PA: PMI.
5. Frame, J. D. (2001) Requirements management: addressing customer needs and avoiding scope creep. In J. Knutson (Ed.), *Project Management for Business Professionals*. New York: Wiley, pp. 63–80.
6. ANSI/EIA-649-1998 (2003) Institute of Configuration Management, Scottsdale, AZ.
7. Cost and value management (1997) *Ibid.*, p. 4.7.3.
8. Cost and value management (1997) *Ibid.*, p. 4.7.3.

9. Globerson, S. (2001) Scope management: do all that you need and just what you need. In: J. Knutson (Ed.), *Project Management for Business Professionals*. New York: Wiley, pp. 49–62.
10. Field, M., and Keller, L. (1998) *Project Management*. London: The Open University, Thomson Learning, pp. 340–341.
11. Field, M., and Keller, L. (1998) *Ibid.*, p. 349.
12. Turner, R. (2008) Managing scope—configuration and work methods. In: R., Turner, (Ed.), *Gower Handbook of Project Management*, 4th ed., Aldershot, UK: Gower, p. 254.
13. Cost and value management (1997) *Ibid.*, pp. 4.7.6 to 4.7.12.
14. Turner, R. (2008) *The Handbook of Project-based Management*, 4th ed., Burr Ridge, IL: McGraw Hill, p. 320.
15. Cost and value management (1997) *Ibid.*, pp. 4.7.12 to 4.7.17.
16. Field, M., and Keller, L. (1998) *Ibid.*, pp. 351–353.
17. Cost and value management (1997) *Ibid.*, p. 4.8.4.

KEY TERMS

Configuration management
Project Management Institute's Body of Knowledge (PMBoK, 4/E)
Systematic management and control of project change
Configuration identification
Configuration status accounting
Configuration audits
Configuration control
Baseline configuration

Chapter 10

Supply Chain Management*

LEARNING OBJECTIVES

- Demonstrate the benefits that accrue to organizations using supply chain management (SCM).
- Discuss the critical areas that supply chain management requires you to focus on.
- Examine the value drivers in a project supply chain.
- Discuss the project supply chain process framework.

During the 1990s, many organizations, both public and private, embraced the discipline of **supply chain management (SCM)**. These organizations adopted several SCM-related concepts, techniques, and strategies to help them gain a significant competitive advantage in the marketplace, including efficient consumer response, continuous replenishment, cycle time reduction, and vendor-managed inventory systems.

Companies that have effectively managed their total supply chain, as opposed to their individual firm, have experienced substantial reductions in inventory- and logistics-related costs, better delivery success, and improvements in customer service. For example, Procter & Gamble estimates that its supply chain initiatives resulted in $65 million savings for its retail customers. The *Journal of Business Strategy* writes that "According to Procter & Gamble, the essence of its approach lies in manufacturers and suppliers

---

Note: Portions of this chapter appeared as Venkataraman, R. (2004) Project supply chain management: optimizing value: the way we manage the total supply chain. In P.W.G. Morris, and J. K. Pinto (Eds.), *The Wiley Guide to Managing Projects.* New York: Wiley.

working closer together jointly creating business plans to eliminate the source of wasteful practices across the entire supply chain."[1]

While the adoption and implementation of total SCM-related strategies is quite prevalent in the retail and manufacturing industries, and their benefits are well understood, project-based organizations have lagged behind in their acceptance and use. This is especially the case with the global engineering and construction industry, which has been plagued by poor quality, low profit margins, and project cost and schedule overruns.[2] It is estimated, for example, that approximately 40 percent of the amount of work in construction consists of non-value-adding activities, such as time spent waiting for approval or for materials to arrive at the project site.[3] Supply chain management clearly offers a better way.

In this chapter, we examine SCM, including its benefits, critical areas, and issues relating specifically to project management. We discuss value drivers in SCM, and present a project supply chain process framework. Finally, we explore supply chain integration, performance metrics, and future issues in project supply chain management.

10.1 WHAT IS SUPPLY CHAIN MANAGEMENT?

Supply chain management is a set of approaches that are utilized to efficiently and fully integrate the network of all organizations and their related activities in producing/completing and delivering a product, a service, or a project. In the process, system-wide costs are minimized while customer-service-level requirements are maintained or exceeded. This definition implies that a supply chain is composed of a sequence of organizations, beginning with basic suppliers of raw materials and extending all the way up to the final customer.

Supply chains are often referred to as **value chains**, as value is added to the product, service, or project as it progresses through the various stages of the chain. Figure 10.1 illustrates typical supply chains for manufacturing and project organizations.

Each organization in the supply chain has two components: inbound and outbound.[5] The inbound component for an organization may be composed of basic raw materials and component suppliers, along with transportation links and warehouses, and ends with the company's internal operations. The outbound component begins where the organization delivers its output to its immediate customer. This portion of the supply chain may include wholesalers, retailers, distribution centers, and transportation companies, and it ends with the final consumer in the chain.

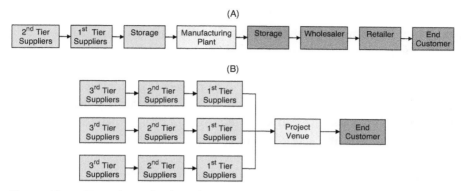

Figure 10.1 Typical supply chain for (A) a make-to-stock manufacturing company and (B) a typical supply chain for a project organization

The length of each component in the supply chain depends on the nature of the organization. For a traditional make-to-stock manufacturing company, the outbound or demand component of the chain is longer than the inbound or supply component. On the other hand, for a project organization, the inbound component is typically longer than the outbound component. These concepts are illustrated in Figure 10.1.

10.2 THE NEED TO MANAGE SUPPLY CHAINS

In the past, businesses focused only on the performance and success of their individual firms. This firm-focused approach, however, does not help companies achieve a competitive edge in the current global business environment, where success (and even survival) increasingly hinges on the ability to manage the total supply chain.

Several factors make it necessary for companies to adopt SCM approaches. First, businesses are encountering competition that is no longer regional or national; it is global, and it is intense. Customers increasingly are seeking the best value for their money, and advancements in information technology and transportation have provided them with the ability to buy from any company anywhere in the world that will provide that value. To win over these customers, business organizations need to reduce costs and add value, not just for their individual firm, but throughout their supply chain.

Second, inventory is a non-value-adding asset and is a significant cost element for businesses. The increasing variability in demand as we move up in the supply chain, known as the "bullwhip effect," can force some individual members of a supply chain to carry very high levels of inventory that can substantially increase the final cost of a product. Effective supply chain management can enable a business to achieve a visible and seamless flow of inventory, thereby reducing inventory-related costs throughout the supply chain.

Third, the chain of organizations involved in producing and delivering a product or completing and delivering a project is becoming increasingly complex and is fraught with many inherent uncertainties. For example, inaccurate forecasts, late deliveries, equipment breakdowns, substandard raw material quality, scope creep, resource constraints, and so on can contribute to significant schedule and cost overruns for a project organization. The more complex the supply chain is, the greater the degree of uncertainty and the more adverse the impact on the supply chain.

Supply chain management approaches such as partnering, information, and risk sharing can greatly reduce the impact of these uncertainties on the supply chain. In addition, management approaches such as **lean production** and **total quality management (TQM)** have enabled many organizations to realize major gains by eliminating waste in terms of time and cost. New opportunities for businesses to improve operations even further now rest largely in the supply chain areas of purchasing, distribution, and logistics.[6]

While several project-based organizations have adopted SCM-related strategies, evidence indicates their efforts to mitigate project schedule and cost overruns have fallen woefully short of expectations. The reason may be that project supply chain management is considerably more difficult, as project supply chains are inherently more complex. For example, many projects typically involve a multitude of suppliers and experience considerable variability in supply delivery lead times and resource constraints, as well as frequent changes to project scope. Such project supply chain complexities underscore the importance and need for project-based organizations to manage their total supply chain in a more formal and organized manner.

10.3   SCM BENEFITS

Companies that effectively manage their supply chain accrue a number of benefits. A recent study by Peter J. Metz of the MIT Center for eBusiness found that companies that manage their total supply chain from suppliers' supplier to customers' customer have achieved enormous payoffs, such as 50 percent reduction in inventories and 40 percent increase in on-time deliveries.[7]

Effective SCM enabled Campbell Soup to double its inventory turnover rate, Hewlett-Packard to reduce its printer supply costs by 75 percent, Sport Obermeyer to double its profits and increase sales by 60 percent in two years, and National Bicycle to achieve an increase in its market share from 5 percent to 29 percent.[8] Other companies that have better managed their supply chain, such as Wal-Mart, have benefited from greater customer loyalty, higher profits, shorter lead times, lower costs, higher productivity, and higher market share. The good news is that by effectively managing their supply chains, project management organizations can enjoy similar benefits.

10.4 CRITICAL AREAS OF SCM

Effective supply chain management requires companies to focus on the following critical areas: customers, suppliers, design and operations, logistics, and inventory.

10.4.1 Customers

First and foremost, effective supply chain management requires a thorough understanding of what customers want—which, in effect, makes them the driving force behind SCM. In a project environment, determining customer requirements and working with them throughout the project will, in all likelihood, lead to a satisfied customer and ultimate success. However, given that customer expectations and needs are constantly changing, determining them can be like hitting a moving target. In recent years, customer value, as opposed to the traditional measures of quality and customer satisfaction, has become more important. According to David Simchi-Levi, Phil Kaminsky, and Edith Simchi-Levi, "Customer value is the measure of a company's contribution to its customer, based on the entire range of products, services, and intangibles that constitute the company's offerings."[9] The challenge for project organizations is to provide this customer value by managing the inevitable scope changes without incurring significant project schedule and cost overruns.

10.4.2 Suppliers

Suppliers constitute the back-end portion of the supply chain and play a key role in adding value to it. Their ability to provide quality raw materials and components when they are needed at reasonable cost can lead to shorter cycle times, reduction in inventory-related costs, and improvement in end-customer service levels.

Traditionally, the relationship between suppliers and buyers in the supply chain has been adversarial, as each was interested in their own profits and made decisions with no regard to their impact on other partners in the chain. In today's environment, supplier partnering is vital for effective supply chain management—without the involvement, cooperation, and integration of upstream suppliers, value optimization in the total supply chain cannot be a reality.

This issue is even more critical for project-based organizations, where the supply or back-end portion of the chain is typically long, and value enhancement and project supply chain performance will be less than optimal without the total involvement of every supplier. For example, in the case of highly technical projects, it is not atypical to have fifth- or even sixth-tier suppliers upstream in the project supply chain.[10] Managing the dynamic interrelationships and interactions that exist among these suppliers is considerably more complex and requires effective integration of their project activities into the larger framework of supply chain management.

10.4.3   Design and Operations

Design and operations play several critical roles in a supply chain. New product designs, for example, often generate new solutions to immensely challenging technical problems. These, in turn, require that changes be made to the existing supply chain (often in the face of uncertain customer demand), as well as possible trade-offs between higher logistics- or inventory-related costs and shorter manufacturing lead times. The operations function creates value by converting raw materials and components into a finished product. This function is present in every phase of the supply chain and is responsible for ensuring quality, reducing waste, and shortening process lead times.

10.4.4   Logistics

Logistics involves the transfer, storage, and handling of materials within a facility, as well as incoming and outgoing shipments of goods and materials. By ensuring that the right amounts of material arrive at the right place and at the right time, the logistics function makes a significant contribution to effective supply chain management. In project management, the logistics function requires a thorough understanding of customer requirements, reduces waste throughout the supply chain to reduce costs, and ensures timely completion and delivery of projects.

10.4.5 Inventory

Inventory control is an essential aspect of effective supply chain management for three reasons. First, inventories represent a substantial portion of the supply chain costs for many companies. Second, the level of inventories at various points in the supply chain has a significant impact on customer service levels. Third, cost trade-off decisions in logistics, such as choosing a mode of transportation, depend on inventory levels and related costs. In project-based organizations, where inventory-related costs can be substantial, effective inventory management can be achieved only through the joint collaboration of all members of the supply chain.

10.5 SCM ISSUES IN PROJECT MANAGEMENT

The benefits of utilizing the total supply chain management approach in traditional make-to-stock manufacturing and retail environments have been well documented. Increasingly, project organizations and project managers are realizing that integrating the total supply chain when managing projects can potentially reduce project schedule and cost overruns, as well as the chances of project failure.

However, as shown in Figure 10.1B, the typical chain for a project is considerably more complex. Problems associated with scope changes, resource constraints, technology, and numerous suppliers that may require global sourcing makes the total integration of the project supply chain risky and challenging.

Consider, for example, the $200 billion Joint Strike Fighter program, one of the most complex project management undertakings in history. According to a 2003 article in *Internet Week*, "The principals of this project supply chain include:

1. A consortium comprised of Lockheed Martin, Northrop Grumman and BAE Systems, overseeing design, engineering, construction, delivery and maintenance,
2. A matrix of partners, including Boeing, engine-makers Pratt & Whitney and Rolls-Royce, and a handful of other subcontractors, all

of which will lean on their own myriad suppliers for hundreds of thousands of components,
3. A multifaceted customer, the Pentagon, which is representing the U.S. Air Force, Navy and Marines, as well as the British Royal Navy and Air Force."[11]

Integrating and managing the total supply chain for this project is a Herculean task that will involve careful balancing of different vested interests and collaboration among all these partners to meet the stringent cost, quality, and delivery criteria set by the Pentagon. If the project's goal is focused only at the department or at the individual company level, instead of on the total project supply chain, value optimization for the project cannot be achieved.

Other issues facing those involved in project supply chain management include

- *Trust among suppliers*—As earlier noted, projects in the construction industry are notorious for ill-managed supply chains. This fact was pointed up in a recent research study of the UK construction sector, which found that fundamental mistrust and skepticism among subcontractors and other supply chain relationships was quite prevalent.[12] Because this lack of trust will have a detrimental effect on the project delivery process, fostering the necessary attitudinal changes to improve performance is a key challenge.
- *Effective inventory management*—In the airline industry, enormous inventory inefficiencies such as duplication of distribution channels and excessive parts in storage have led to increasing costs for the total supply chain. In addition, a significant portion of every dollar invested in spare parts inventory constitutes holding and material management costs. The challenge for airlines and other project-based industries is to efficiently manage inventory throughout the total project supply chain, which in turn creates the potential to significantly reduce project life cycle costs.
- *Joint coordination of activities and communication among various project participants*—Consider, for example, a development project for an aluminum part to be delivered to an airline customer. When the part is ready for production, the supply chain department of the project development group will typically choose from a list of its favorite suppliers to get the lowest possible price. These companies are rarely the ones that worked on the development hardware, and not surprisingly, they

will all want to enforce design changes on engineering, so that they can efficiently produce the part to fit their particular set of processes. This can often lead to substantial increases in costs by way of engineering modification and re-qualification efforts. Furthermore, if, in the interest of price reduction, the supply chain department later changes the design without communicating or coordinating with the engineering department, the part that will be delivered to the customer will be different from what the customer wanted. Much time, money, and effort may have to be expended to rectify the situation with the irate customer. Project managers should be aware that without the joint collaboration of all project stakeholders working toward a common goal, suboptimization will occur, and the project is bound to fail.

- *Accurate, timely, and quality information on supply-chain-related issues*—Often, this information is not available to project managers—resulting in suboptimal decisions. Effective project supply chain management requires an infrastructure that can accelerate the velocity of information and will enable all project participants to collaborate throughout the project life cycle. For example, in a chemical plant construction project, the Global Project and Procurement Network uses the Internet to streamline and accelerate information flow. This enables all supply chain participants to collaborate from plant design through operation and maintenance.[13]
- *Security management*—For many organizations, the terrorist attacks of September 11, 2001 have heightened interest in ensuring security throughout the total supply chain. The challenges—which range from designing facilities that are secure against outside intrusion to ensuring that products can be protected from tampering until they reach the end consumer—require that project managers provide unique and innovative solutions.

10.6 VALUE DRIVERS IN PROJECT SUPPLY CHAIN MANAGEMENT

Value drivers in a project supply chain are those strategic factors that significantly add or enhance value and provide a distinct competitive advantage to the chain. The typical value drivers for a project supply chain are listed in Table 10.1.

In the context of project supply chains, the client or **customer** who is the final recipient of the completed project is the most important value

Table 10.1 Project supply chain value drivers

Value drivers	Definition
Customer	The final customer at the end of the project supply chain
Cost	Total cost incurred at the end of the project supply chain
Flexibility	The ability of the project supply chain to quickly recognize and respond to changing customer needs
Time	Refers to on-time delivery or delivery speed of completed projects to the end customer
Quality	The ability to deliver a completed project that meets or exceeds end customer expectations

driver. It is this customer's definition or perception that determines what constitutes value in a project, which then triggers all other upstream supply chain activities. For example, if the customer values price, then all supply-chain-related activities of the project should focus on efficiency and eliminating waste throughout the total supply chain. On the other hand, if the customer values completion of the project on time or ahead of schedule, then all of the project supply chain activities should be geared toward achieving this goal. Thinking in terms of customer value requires that project managers have a clear understanding of customer preferences and needs, customer profit and revenue growth potential, and the type of supply chain required to serve the customer. They must also make sure the inevitable trade-offs that need to be made are indeed the correct ones.[14]

The need to significantly lower or control project **costs** will also drive changes and improvements in the supply chain. In the retail industry, for example, the policy of everyday low prices required Wal-Mart to adopt a cross-docking strategy in its warehouses and distribution centers, as well as strategic partnering with its suppliers. In the personal computer industry, Dell Computer Corporation uses the strategy of postponement (i.e., delaying final product assembly until after the receipt of the customer order) to lower its supply chain costs.

Flexibility, or the ability to respond quickly to changes in customer needs or project scope, is another important value driver in project supply chains. For example, the willingness of the project organization to provide the client with the freedom to make significant design changes through development, with the help of a strong and supportive engineering staff, will enhance the value of the project supply chain.[15] Dell Computer Corporation is a classic example of a company that used flexibility to enhance customer perception of value. By allowing the customers to configure their own personal computer systems, Dell gained a significant competitive advantage in its industry.

The dimension of **time** has always been an important success factor in project management. Time, in the form of project scheduling and in conjunction with cost and quality, represents the three most important constraints in projects. In event project management, such as the Olympic Games, the dimension of time is of overriding importance, as the whole world is watching and the games must start on schedule. In other project-oriented situations, however, cost or quality can be more important value drivers, and trade-offs in terms of time may have to be made in such projects. In any event, the ability to complete a project on time or ahead of schedule will certainly contribute to value in project supply chain management. In the retail industry, for instance, several time-based supply chain strategies such as continuous replenishment systems, quick response systems, and efficient consumer response evolved as a direct result of the value-adding nature of time.

Quality, in a project context, is defined as achieving the project objectives that are "fit for purpose." *The Handbook of Project-based Management* states that, "Fit for purpose means that the facility, when commissioned, produces a product which solves the problem, or exploits the opportunity intended, or better. It works for the purpose for which it was intended."[16]

Project quality, simply defined, is that the project's product meets or exceeds customer expectations.[17] Quality has several dimensions. For example, a person wanting to buy a Steinway grand piano for a price of $25,000 is more likely interested in the performance dimension of quality, whereas a person who wants to buy a Baldwin vertical piano for $5000 is probably looking for a piano of consistent quality.

Understanding the level of quality a customer wants in a project, ensuring the functionality of the project's product at that level of quality, and delivering the project at a reasonable price and time that will delight the customer should be the ultimate goal of every project manager. Meeting or exceeding the quality expectations of the customer adds value by fostering and sustaining customer loyalty and goodwill long after the project is completed.

Achieving this level of quality in projects, however, is easier said than done. It requires the total commitment to quality by every member of the total project supply chain, along with the integration of all quality management activities. A way to achieve this commitment is to adopt and integrate total quality management (TQM) in project supply chains.

10.7 OPTIMIZING VALUE IN PROJECT SUPPLY CHAINS

Setting up the optimal supply chain for any project-based organization is an intricate and multilayered challenge. The critical feature to keep in mind is that our goal is not some arbitrary or vague desire to "enhance" the supply chain; it is a direct effort to improve the value that our organization can derive from these external, but critical, linkages. Thus, there are several elements to increasing value in project supply chains identified below.

10.7.1 Total Quality Management

While this topic is covered much more thoroughly in Chapter 11, it does bear special mention here as part of the project supply chain management process. As defined by J. Rampey and Harry V. Roberts, "TQM is a holistic approach to continuously meeting customer needs and aims at continual increase in customer satisfaction at continually lower real cost. Total Quality is a total systems approach (not a separate area or program), and an integral part of high-level strategy. It works horizontally across functions and departments, involving all employees, top to bottom, and extends backwards and forwards to include the supply chain and customer chain."[19]

The integration of the project supply chain's quality management activities through TQM is vital to complete a project in such a way that the multiple objectives of the customer can be met in terms cost, quality, time, and safety. The construction industry, for instance, is increasingly embracing TQM to solve its quality problems and ensure customer satisfaction. Quality assurance has always been difficult in this industry, as products are one-off, production processes are nonstandardized, and project design changes are frequent. Furthermore, the general contractor for a construction project is totally dependent on the goods and services of suppliers and other subcontractors to meet the quality requirements of the customer.

When quality management activities in a construction project supply chain are integrated through the application of TQM, general contractors, suppliers, and other subcontractors are able to improve their own quality performance, and can contribute toward optimizing customer value. For example, since the Shui On Construction Company in Hong Kong successfully adopted TQM in 1993, it has been known for its good performance in building housing projects, and won the "Contractor of the Year" award three years in a row from 1995 to 1997.[20]

10.7.2 Choosing the Right Supply Chain

A fundamental prerequisite for value optimization in projects is the choice of the right supply chain for the project. More often than not, less-than-stellar supply chain performance is due to the mismatch between the nature of the product and the type of supply chain chosen to produce it.[18] In the context of a project environment, this implies that first and foremost, the nature of the project should be clearly delineated—whether it is primarily functional or primarily innovative. The next step is to choose the right supply chain for this project that will directly contribute to its core competencies and provide a distinct competitive advantage. Without having the right supply chain that is best for a particular project, value optimization in projects cannot be achieved.

10.8 PROJECT SUPPLY CHAIN PROCESS FRAMEWORK

The rest of the discussion in this section is based on a simple framework of the project supply chain process that is presented in Figure 10.2. In this figure, the square box represents the procurement component of the chain. The oval box represents the conversion or fabrication phase of the project, where the product is created, and the rectangular box represents delivery of the completed project to the customer.

10.8.1 Procurement

Procurement involves all activities that are vital in acquiring goods or services that will enable an organization to produce the product or complete a project for its customer. The procurement portion of the project supply chain is typically long, and it is not uncommon to find fifth- or even sixth-tier suppliers upstream. This is the area where the greatest opportunities for cost reduction and enhancing value of the total project supply chain exist.

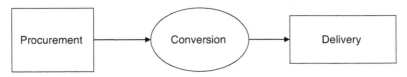

Figure 10.2 Project Supply Chain Process Framework

The decision to buy goods or services from an outside vendor should be made only after a thorough "make or buy" analysis that determines whether producing a product or component in-house will directly contribute to the organization's set of core competencies. If it will not, the product or components should be purchased from outside suppliers.

When the decision is made to "go outside," procurement involves identifying and analyzing user requirements and type of purchase, selecting suppliers, negotiating contracts, acting as liaison between the supplier and the user, and evaluating and forging strategic alliances with suppliers. For many organizations, materials and components purchased from outside vendors represent a substantial portion of the cost of the end product. In fact, companies such as Boeing and Rolls-Royce typically incur 60 percent of their project cost and 70 percent of their lead time because of purchased materials.

In these situations, where effective procurement can significantly enhance an organization's competitive advantage, managing suppliers and ensuring that parts and components of appropriate quality are delivered on time is a truly daunting challenge. Examples of project-based industries where this is particularly true include aerospace and construction.

In the aerospace industry, effective procurement strategies—including international sourcing, long-term supplier contracts, partnering with suppliers in project design, and risk and information sharing—can maximize a company's purchasing power, contribute to its business success, and significantly enhance the value of its supply chain.

This was learned the hard way by Boeing in 1997, when the company attempted to double its production overnight to respond to an unprecedented demand for new airplanes—without realizing the impact the move would have on its supply chain. The move resulted in parts and worker shortages at the assembly stage that forced Boeing to close its 747 and 737 assembly lines, and the company was hit with a $1.6 billion loss. Four years later, Boeing began to revamp its supply chain process through the use of lean manufacturing techniques, and now requires tighter integration with suppliers and just-in-time delivery of parts.[21]

On the other hand, an example of successful procurement strategies was that of Sikorsky Aircraft, which in 1993 adopted the method of Supplier Kaizen. This method, described as "bringing the suppliers to the same level of operations as the parent company, through training and improvement projects, to ensure superior performance and nurture the trust that is required for strong partnerships,"[23] yielded tremendous benefits to Sikorsky Aircraft. As a result of applying Supplier Kaizen to its procurement practices, the company achieved long-term commitments from suppliers to partner for future growth, declining prices, and shorter lead times.[22]

In the construction industry, which as a whole is characterized by mutual distrust and antagonism within existing supply chain relationships, the key to future performance improvement is through the adoption of effective procurement strategies. These include supplier selection and partnering, e-procurement, and Supplier Kaizen.

In fact, a recent Hong Kong-based study of factors affecting the performance of construction industry has shown that the methods used for selecting the overall procurement system, contractors, and subcontractors are critical, and the use of information technology/information systems can facilitate appropriate selection through all stages of the construction supply chain.[24]

In the context of supplier selection, the series of international standards on quality management and quality assurance developed by the International Organization for Standardization (ISO) and called ISO 9000 can be highly useful. For instance, companies that are ISO 9001 certified have demonstrated to an independent auditor that their systems and operations have met rigorous international standards for quality, and therefore can be included in lists of potential suppliers.

Supply Chain Relationships

Value optimization in projects cannot be achieved in the absence of close and trusting relationships among project participants. Building trust and integrating information systems among supply chain members can lead to the elimination of certain redundant processes and simplification of sourcing, negotiating, and contracting procedures. Planning efficiency and project performance will improve, because the availability of timely and accurate information puts suppliers in a better position to provide valuable input to project planning.[25]

A recent study of two construction projects in the United Kingdom has shown that significant supply chain benefits and improvements can be realized through close partnerships and involvement of suppliers and subcontractors very early on in the project.[26] When all members of the supply chain are involved in translating the design concept into reality, they're better able to ensure that appropriate cost criteria are met. Suppliers, along with other partners, can be more innovative; problems can be resolved early, as there are more open channels of communication; and the end result will be a project that is completed on time, of higher quality, at a lower cost, and that provides clients with better value for their money. It is estimated that in the construction industry, supply chain partnering alone would lead to a 10 percent reduction in cost and time, similar increases in productivity and quality, and a 20 percent reduction in defects and accidents.[27]

Supplier Development

Supplier development is yet another strategy that can add value to the procurement phase of the project supply chain. General Electric Company, as part of their global sourcing initiative, has a program for supplier development in which GE personnel provide extensive training to vendors in improving their own operations. Vendors who attain to the level of quality and efficiency that GE requires are awarded long-term contracts. In the final analysis, procurement in a project context requires extensive planning and coordination of project activities with suppliers. Strategies such as Supplier Kaizen, partnering based on trust, vendor development, information and risk sharing, long-term strategic alliances with suppliers, and integrating quality management activities of suppliers through TQM will significantly reduce procurement and inventory costs, shorten lead times, and improve quality of purchased materials, and enhance the value of the total project supply chain.

10.8.2 Conversion

The next phase of the project supply chain (shown in Figure 10.2) that requires attention for value optimization is the conversion, or fabrication, phase. This is the venue where the project's product is actually created, as in the case of new product development, creation of a new software package, or building an offshore oil-drilling vessel.[28] To a large extent, the degree of successful value that can be achieved in this area is dictated by the efficiency and effectiveness of the procurement phase.

As in the case of procurement, the challenges encountered in this phase will depend on whether the project is relatively routine or highly complex. Regardless of the nature of the project, however, several strategies that have proven to be successful in the traditional manufacturing environment can be employed to enhance value in the conversion phase. For example, the application of lean manufacturing techniques can add value by eliminating waste and unnecessary inventories, and by shortening process lead times.

This is the strategy of Boeing Corporation, which is moving to thwart stiff competition from Airbus by employing lean manufacturing practices. Their goal is to create an innovative, company-wide implementation of gigantic, moving assembly lines in the commercial aircraft division—which is guaranteed to speed up production by 50 percent and increase profit margins to double-digit levels on commercial aircraft sales.[29]

10.8.3 Delivery

The final phase of the project supply chain process in Figure 10.2 is delivery of the completed project to the customer. Normally, the transfer of the completed project is relatively straightforward. In recent years, however, the project delivery process has undergone some significant changes, particularly in the case of clients from foreign countries. For example, in large plant-construction projects, some foreign countries require the project organization to operate the plant jointly with the foreign client for some extended period of time to mitigate potential start-up problems and reduce client risk.[30] With clients becoming increasingly risk-averse, the willingness of the project organization to assume some additional risks is certain to add value and provide a distinct competitive advantage to the total project supply chain.

10.9 INTEGRATING THE SUPPLY CHAIN

The obvious key to value optimization in projects is the total integration of various components of the project supply chain. Several strategies can be implemented to achieve this goal. First, as shown by a recent study of two demonstration projects in the United Kingdom, the development of "work clusters" and the application of concurrent engineering principles can lead to project supply chain integration, which in turn can improve value, eliminate inefficiencies, and reduce project costs.[31]

Second, project supply chain integration can be achieved through collaboration and standardization of business processes among project supply chain partners. Such collaboration, however, requires an understanding and management of the differences and interests of all project supply chain members to create a common vision and work culture.[32]

Third, accelerating information velocity by building an Internet-based supplier network for procurement purposes can further facilitate collaboration and integration in project supply chains.[33] Building such a network also presupposes the presence of a viable IT/IS infrastructure among project supply chain members. For example, the Joint Strike Fighter project discussed earlier in this chapter will require that the various organizations involved in design, engineering, manufacturing, logistics, finance, etc. collaborate over the Internet to meet the stringent cost, quality, and time requirements set by their customer, the Pentagon.

Fourth, project supply chain integration and the resulting value optimization require that supply chain partners change traditional thinking and practices. Effecting such a change requires the commitment and involvement of the people in each organization in the project supply chain, and should begin with the senior management of each partner. Ultimately, it is the responsibility of senior managers to prepare their organization for change, to overcome cultural and organizational barriers to change, and to achieve cross-functional and cross-business unit cooperation.[34] Without this, project supply chain integration and value optimization cannot be achieved.

In addition to these strategic initiatives, the following practical steps can be undertaken to add value to a project[35]:

1. *Flowchart the project supply chain processes before the project is initiated*—This process will show the various links or steps involved in completing the project, and each step will potentially have a customer and a supplier. The flowchart can identify potential areas of redundancies, waste, or other non-value-adding activities in the chain, and can facilitate the use of lean management initiatives to eliminate them.
2. *Standardize processes*—Standardization of processes throughout the project supply chain by the use of methods such as simultaneous design, concurrent engineering, lean manufacturing, mistake proofing, total productivity maintenance, and collaborative teamwork will ensure consistency.

3. *Control process variation*—It is essential that processes across the total project supply chain are monitored and controlled for variation, including lead times, quality in materials, and production processes. Once the supply chain processes are stabilized, they can be improved.
4. *Prequalify suppliers through supplier certification*—Ensure that suppliers in each link of the project supply chain process are QS-9000 or ISO 9001 certified. This certification guarantees a pool of quality suppliers.
5. *Audit project supply chain processes and take corrective and preventive actions*—Processes should be audited periodically for improvement and risk identification. Corrective action should be taken to eliminate the root causes of nonconformance and deficiencies that were uncovered through the audit, while preventive action will ensure that these problems do not reoccur.
6. *Measure project supply chain performance*—Without the availability of specific quantifiable performance metrics, project supply chain performance in terms of both efficiency and customer satisfaction cannot be gauged. For this reason, performance metrics should be developed and used, and competitive benchmarking should be performed. These tools will convey immediately how the project supply chain has been performing over time or in comparison with best-in-class competitors.

10.10 PERFORMANCE METRICS IN PROJECT SUPPLY CHAIN MANAGEMENT

Measuring project supply chain performance is a complex and challenging endeavor. Appropriate, meaningful metrics should be carefully developed at the planning stage, when the total project supply chain is being designed, and all members of the project supply chain should be involved. This will require reconciling differences and reaching consensus regarding appropriate metrics to measure success, as compared to best-in-class competitors. While there are a number of ways to classify performance metrics, we will examine project supply chain performance using the project process value drivers of time, cost, quality, and flexibility. These performance metrics categories are presented in Table 10.2.[36]

Time—in particular, project completion time—has always been considered an important measure of project performance. However, this metric

Table 10.2 Project supply chain performance metrics categories

Performance category	Performance issues
Time	1. Was the project completed and delivered on time? 2. What is the potential variability in project completion times? 3. Was the completed project operationalized on time to the satisfaction of the customer? 4. Were the purchased materials and manufactured components delivered on time by upstream suppliers? 5. What is the potential variability in procurement lead times?
Cost	1. Was the completed project within budget for each of the project supply chain members? 2. What was the total project supply chain cost? • Procurement cost of purchased materials • Manufacturing cost • Inventory-related cost • Transportation cost • Project acceleration costs • Cost of liquidated damages • Other relevant costs: administrative, etc.
Quality	1. Did the project meet the technical specifications and does it provide the functionality desired by the customer? 2. Was the customer satisfied with the service provided during start-up, implementation, and final project transfer? 3. Were the purchased raw materials and manufactured components defect-free? 4. Was the completed project's product reliable and durable during its life cycle?
Flexibility	1. Was the customer accorded reasonable freedom within a reasonable timeframe to make changes to the project scope, design, or specifications? 2. Were upstream suppliers responsive to the reasonable needs of their downstream partners in terms of delivery time and quality issues?

Performance Metrics in Project Supply Chain Management 229

should also capture other elements of time, such as operational and start-up times, and procurement and manufacturing lead times. Furthermore, for routine projects, the potential variability in these times should also be measured to track consistency and reliability of the project supply chain. For example, assume that, historically, the estimated completion time for routine construction projects has been 36 weeks. How frequently the project supply chain achieves this completion time is an indicator of consistency and reliability, and can provide important insights for future supply chain improvements.

Some of the cost metrics noted in Table 10.2 are fairly straightforward. The important caveat here is that the emphasis should be on the cost incurred for the total project supply chain, and not just the cost incurred by the project organization. The total project supply chain cost is multidimensional and includes several elements, such as procurement and manufacturing cost of materials and goods, inventory costs, and so forth. Focusing on the total cost incurred will enable project participants to identify inefficiencies in the supply chain and facilitate coordination to devise ways to eliminate them. The ultimate goal is to optimize value by reducing waste and unnecessary cost throughout the supply chain.

Like cost, the quality metric also has several dimensions. In a project context, the most obvious ones are the dimensions of performance—that is, the functionality of the project's product and conformance to design or technical specifications. In addition to these dimensions, the level of service provided to the customer during the start-up and implementation phase and throughout the project's life cycle are also important quality measures. Ultimately, it is the customer's perception of quality that matters, and the response of the project supply chain to meet this value perception should be the focus of this metric.

The last project supply chain performance metric category is flexibility. This metric measures the willingness and ability of the project supply chain to respond to reasonable changes in scope or design requested by the customer. Building an effective configuration and change control system that spans the total project supply chain can help achieve such flexibility and provide a distinct competitive edge to the value chain.

10.11 PROJECT SUPPLY CHAIN METRICS AND THE SUPPLY CHAIN OPERATIONS REFERENCE (SCOR) MODEL

Project supply chain metrics span the entire supply chain, with specific focus on common processes, and should capture all aspects of supply chain performance. The SCOR model developed by the Supply Chain Council (SCC) provides the framework to track such performance, and has been the basis for supply chain improvement for both global and site-specific projects.[37] By integrating the well-known concepts of business process reengineering, benchmarking, and process measurement, the SCOR model provides a cross-functional framework for improving supply chain performance. It spans all aspects and interactions of the supply chain, from the customer's customer all the way back to the supplier's supplier. According to *The Management of Business Logistics: A Supply Chain Perspective*, "The SCOR model provides standard descriptions of relevant management processes; a framework of the relationships among the standard processes; standard process performance metrics; and standard alignment to features and functionality. The ultimate aim is to produce best-in-class supply chain performance."[38]

The model uses five key aspects of a supply chain—plan, source, make, deliver, and return—as building blocks to describe any supply chain. The SCOR model can be adapted to describe a project supply chain, as shown in Figure 10.3.

In a project supply chain context, the process outlined in Figure 10.3 encompasses all aspects of planning in an overall project supply chain

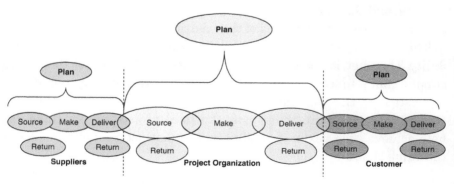

Figure 10.3 The SCOR model adapted for a project supply chain
Source: Supply Chain Council.

plan, including the integration of the individual plans of all supply chain members.

- The **planning** phase essentially involves understanding customer needs and project scope; determining the best course of action to meet the sourcing, producing, and delivery requirements of the project; and developing the criteria to evaluate total project supply chain performance.
- The **sourcing** phase focuses on all processes related to procurement, such as identifying, selecting, and qualifying suppliers, negotiating contracts, managing inventory, and so on.
- The **"make"** process encompasses all aspects of creating the project's product, such as design, testing, building, and completing the project. It also includes systems and processes for quality and change control and performance reporting.
- The **delivery** process covers all aspects related to the final transfer of the completed project to the customer, including installation, start-up, etc. to the satisfaction of the customer.
- The **return** phase encompasses activities that can range from addressing problems associated with the completed project's functionality at the customer site to return of raw materials to the vendor.

The SCOR model for a project supply chain is composed of three levels. At the top level, the scope and content of the model is defined, and performance targets based on best-in-class competition are established. The next level focuses on the configuration of project's supply chain, and the last level includes process elements such as performance metrics, systems and tools, best practices, and the system capabilities to support them. Because the SCOR model is based on standard processes and standard language, meaningful performance metrics for the project supply chain can be developed.

10.12 FUTURE ISSUES IN PROJECT SUPPLY CHAIN MANAGEMENT

Project supply chains in the twenty-first century will encounter a number of challenges, as well as opportunities for improvement. First, the availability and power of information technology will drastically transform project supply chains, facilitating virtual integration and providing the benefits that accrue from tight coordination, partnering, quick and efficient communication, focus, and specialization. At the same time, the Internet and related e-commerce technologies can be exploited to overcome major systemic constraints. The challenge will be to create and build a boundary-spanning information infrastructure that enables quick and efficient information sharing and communication.

Second, the trend toward globalization in project supply chain management will accelerate, as it has the potential to provide significant cost advantages. One example of this is Boeing Corporation, which has partnered with European suppliers to procure higher-level assemblies as part of an initiative to reduce the number parts handled in its production lines.[39]

Finally, businesses are becoming increasingly concerned about the environment and are undertaking environmental projects to reduce costs, reduce pollution and hazardous materials, improve manufacturing performance and quality, improve relationships with external stakeholders, and proactively deal with environmental regulations. This trend toward environmental friendliness will require supply chains to address issues like recycling, reuse, asset recovery, minimization of waste, and handling and disposal of hazardous materials.[40]

To effectively respond to these challenges and exploit the opportunities, project supply chains need to adopt a comprehensive and integrated supply chain perspective, which will significantly enhance their value and create a distinct competitive advantage.

REFERENCES

1. Cottrill, K. (1997) Reforging the supply chain. *Journal of Business Strategy*, 18 (6): 35–39.
2. Yeo, K. T., and Ning, J. H. (2002) Integrating supply chain and critical chain concepts in engineer-procure-construct (EPC) projects. *International Journal of Project Management*, 20: 253–262.
3. Mohamed, S. (1996) Options for applying BPR in the Australian construction industry. *International Journal of Project Management*, 14 (6): 379–385.
4. Watson, K. (2001) Building on shaky foundations. *Supply Management*, 6 (17): 22–26.

5. Stevenson, W. J. (2002) *Operations Management*, 7th ed. New York: McGraw-Hill/Irwin.
6. Stevenson, W. J. (2002) *Ibid*.
7. Betts, M. (2001) Kinks in the chain. *Computerworld*, 35 (51): 34–35.
8. Stevenson, W. J. (2002) *Ibid.*; Fisher, M. L. (1997) What is the right supply chain for your product? *Harvard Business Review*, 75 (2): 105–116.
9. Simchi-Levi, D., Kaminsky, P., and Simchi-Levi, E. (2003) *Designing and Managing the Supply Chain: Concepts, Strategies and Case Studies*, 2nd ed. New York: McGraw-Hill/Irwin, p. 11.
10. Pinto, J. K., and Rouhiainen, P. J. (2001) *Building Customer-based Project Organizations*. New York: Wiley.
11. Preston, R. (2001) A glimpse into the future of supply chains. *Internet Week*, 885: 9–10.
12. Dainty, A.R.J., Brisco, G. H., and Millet, S. J. (2001) Subcontractor perspectives on supply chain alliances. *Construction Management and Economics*, 19 (8): 841–848.
13. Cotrill, K. (2001) Engineering a value chain. *Traffic World*, 265 (9): 21–22.
14. Simchi-Levi, D., Kaminsky, P., and Simchi-Levi, E. (2003) *Designing and Managing the Supply Chain: Concepts, Strategies and Case Studies*, 2nd ed. New York: McGraw-Hill/Irwin.
15. Pinto, J. K., and Rouhiainen, P. J. (2001) *Ibid*.
16. Turner, J. R. (2008) *The Handbook of Project-based Management*. Berkshire, UK: McGraw-Hill, p. 150.
17. Turner, J. R. (2008) *Ibid*.
18. Fisher, M. L. (1997) *Ibid*.
19. Rampey, J., and Roberts, H. V. (1992) Perspectives on total quality. *Proceedings of Total Quality Forum IV*, Cincinnati, Ohio.
20. Wong, A., and Fung, P. (1999) Total quality management in the construction industry in Hong Kong: a supply chain management perspective. *Total Quality Management*, 10 (2): 199–208.
21. Holmes, S. (2001) Boeing goes lean. *Business Week*, June 4: 94B–94F.
22. Foreman, C. R., and Vargas, D. H. (1999) Affecting the value chain through supplier Kaizen. *Hospital Materiel Management Quarterly*, 20 (3): 21–27.
23. Foreman, C. R., and Vargas, D. H. (1999) *Ibid*.
24. Kumaraswamy, M., Palaneeswaran, E., and Humphreys, P. (2000) Selection matters: in construction supply chain optimization. *International Journal of Physical Distribution and Logistics Management*, 30 (7/8): 661–669.
25. Yeo, K. T., and Ning, J. H. (2002) *Ibid*.

26. Ballard, R., and Cuckow, H. J. (2001) Logistics in the UK construction industry. *Logistics and Transportation Focus*, 3 (3): 43–50.
27. Watson, K. (2001) *Ibid*.
28. Pinto, J. K., and Rouhiainen, P. J. (2001) *Ibid*.
29. Holmes, S. (2001) *Ibid*.
30. Pinto, J. K., and Rouhiainen, P. J. (2001) *Ibid*.
31. Nicolini, D., Holti, R., and Smalley, M. (2001) Integrating project activities: the theory and practice of managing the supply chain through clusters. *Construction Management and Economics*, 19 (1): 37–47.
32. Padhye, A. (2001) Apply leverage to ensure business process integration in your supply chain. *EBN*, 1282: L38.
33. Cotrill, K. (2001) *Ibid*.
34. Burnell, J. (1999) Change management is the key to supply chain management success. *Automatic I. D. News*, 15 (4): 40–41.
35. Hutchins, G. (2002) Supply chain management: a new opportunity. *Quality Progress*, 35 (4): 111–113.
36. Coyle, J. J., Bardi, E. J., and Langley, C. J. (2010) *The Management of Business Logistics: A Supply Chain Perspective*, 8th ed. Cincinnati; OH: South-Western.
37. Yeo, K. T., and Ning, J. H. (2002) *Ibid*.
38. Coyle, J. J., Bardi, E. J., and Langley, C. J. (2010) *Ibid*.
39. Sutton, O., and Cook, N. (2001) Quest for the ideal supply chain. *Interavia*, 56 (657): 24–27.
40. Carter, C. R., and Dresner, M. (2001) Purchasing's role in environmental management: cross-functional deployment of grounded theory. *Journal of Supply Chain Management*, 37(3): 12–26.

KEY TERMS

Supply chain management (SCM)
Value chains
Lean production
Total quality management (TCM)
Logistics
Customer
Flexibility

Time
Quality
Procurement
Conversion
Delivery
SCOR model

Chapter 11

Quality Management in Projects

LEARNING OBJECTIVES

- Examine the elements of project quality.
- Explain the definition of quality engineering.
- Illustrate the key principles of Total Quality Management in projects.
- Demonstrate the application of six-sigma methodology for projects.

In terms of enhancing value in projects, cost, time, and quality are the three key criteria. However, the term "quality" can mean different things to different stakeholders, and there is no single definition that is universally applicable. This is also true in a project environment. Because project managers and other project stakeholders are uncertain about what constitutes quality in a project context, there is greater uncertainty among them when asked about the process required to manage quality. Many project participants refer to a project's functionality and performance as the key dimensions of quality. While these definitions are certainly valid, they are also narrow. When other factors are considered, including the service expected by the customer during and after installation, the ease of use of the project's facility, and so on, a much broader definition of quality is warranted.

In this chapter, we take this expanded view of quality and examine the various elements of managing quality in a project environment. We also take an in-depth look at the concept of total quality management, as well as quality management methods for a project organization. Finally, we examine the Six Sigma model for projects, including its practical application.

11.1 DEFINITION OF QUALITY IN PROJECTS

Managing quality in projects must be addressed from two different perspectives: the quality of the product of the project, and the project quality management process. Issues associated with product quality, such as quality metrics and required tools and techniques, are very specific to the nature of the product. For example, the quality issues to be addressed and approaches to be used in building a convention center will be significantly different from those of manufacturing a jet engine. On the other hand, the project quality management process is applicable to a whole spectrum of projects, with wide variation in the nature of the product from project to project. It includes all necessary activities undertaken by the project organization to ensure that the needs of the project and the purpose for which it was initiated are fully met, such as determining quality policies, objectives, and responsibilities.[1]

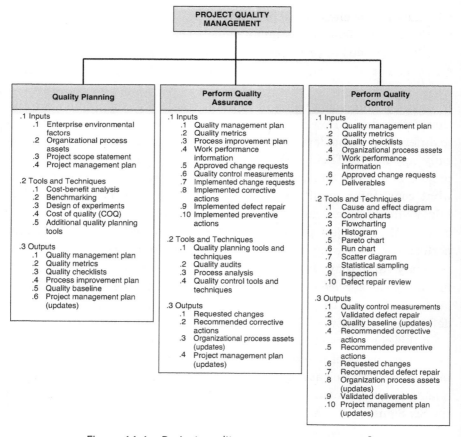

Figure 11.1 Project quality management overview[3]

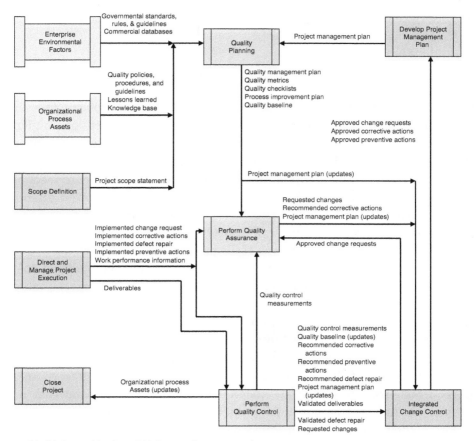

Figure 11.2 Project quality management process flow diagram[4]

The project quality management process facilitates the implementation of a quality management system through policies, procedures, and the subprocesses of quality planning, quality assurance, and quality control. An overview of the project management process is presented in Figure 11.1, while the various inputs, outputs, and subprocesses that constitute the overall project quality management process are presented in the form of a process flow diagram in Figure 11.2.[2]

11.2 ELEMENTS OF PROJECT QUALITY

There are six elements that impact project quality: the project's product (outcome), management processes, quality planning, quality assurance, quality control, and corporate culture. In the following section, each of these is discussed in detail.

11.2.1 The Project's Product

The project's product is expected to solve the customer's problem, provide required functionality, and generate revenue for the project sponsor. For these reasons, the quality of the end product is a vital determinant of overall project quality. Broadly defined, **product quality** in a project context refers to the ability of the project's facility or product to meet or exceed customer expectations. In a project environment, there is often a gap between the customer's definition of *project* quality and the project organization, project manager, and project team's definition of *product* quality. This is because the focus of the project team is often on designing functionality, and determining whether or not that functionality is met when the project is commissioned. The customers' view of project quality centers around problems that they want solved through the project, which eventually constitute their requirements. The task of the project organization is to translate these requirements into design specifications and deliver the project in accordance with them.

Assuming that all of the necessary steps required to translate customer requirements into design specifications have been taken, and that no mistakes have been made in the project's construction, delivery, or installation, the project will fit the purpose for which it is intended. While this sounds simple enough, it is easier said than done. The reason why it is difficult to bridge the gap between quality expectations is because all the necessary steps required to translate customer requirements are often *not* taken, and mistakes *are* made. In addition, the customer is often not sure of what the exact requirements are—which makes it very difficult to translate them into design specifications. Consequently, it is unlikely that the project's facility or product will solve the customer's problem or be fit for its purpose, let alone provide customer satisfaction.

For these reasons, the appropriate definition of project quality originates from the customer; in other words, "the project's product or facility is fit for its intended purpose, solves the customer's problems, and its performance satisfies or exceeds customer expectations."[5] This definition implies that the customer's requirements definition (scope) and subsequent design specifications can be imperfect, and will need refinement.

Imperfect requirement definitions lead to inevitable scope changes, which in turn lead to subsequent changes to design specifications. This is often the biggest source of conflict between the customer and the project sponsor, because the sponsor is often reluctant to change design specifications after time, effort, and money have already been expended. However, if the specifications are imperfect, the project's product will not fit the purpose for which it is intended, and should be changed.

While changes to specifications may become necessary, they should be infrequent, should be used sparingly, and should be managed in a formal manner using the configuration management system discussed in Chapter 8. As noted in Figure 11.1, issues such as proper scope definition and managing scope changes fall under the purview of the project quality management process.

The ultimate quality of the product delivered begins with the quality of design, or the designers' intention to include or exclude certain features of the project's product. The decision on quality of design is the result of a collaborative effort between the customer and the project design team. The emphasis is to ensure that all of the features required to provide the desired functionality of the product are incorporated. The design decision should also factor into account the operational capabilities of the project organization, costs, and safety and liability issues—both during the project development stage, and after delivery of the completed project.

It should be noted that a poor initial design can create considerable problems during the project development phase, including difficulty in acquiring required materials, or inability to meet specifications or follow procedures. Ultimately, poor quality of design will result in project cost overruns and lower project value. One approach that project designers can use is quality engineering, which originated in the late 1980s by Japanese engineer Genichi Taguchi. While a detailed discussion of the quality engineering approach is beyond the scope of this book, the next section provides a brief overview.

Quality Engineering

The **quality engineering** approach involves a combination of engineering and statistical methods that can reduce costs and improve product quality by optimizing the design of a product and its production processes. Taguchi argued that the quality of a product should not be perceived as whether or not it is within or outside of specifications; instead, it should viewed in terms of variation from the quality characteristic's target value. Any variation from this target value is a loss in quality, with associated costs that are undesirable and unwelcome. In addition, Taguchi believed that managers should actively seek ways to reduce all variations from the target value, instead of being satisfied when a product's quality characteristic value is within specified limits.

Taguchi devised a mathematical formula known as the **quality loss function** for determining the costs associated with poor quality. He contended that when the product's quality characteristic is exactly on the target value, the quality loss function has a value of zero. The value of this loss function increases exponentially as the value of the quality characteristic of the product approaches the specification or tolerance limits. Clearly, when the quality of a product barely conforms to specifications, it is defective, especially when compared to one whose quality characteristic value is right on target. A schematic representation of Taguchi's quality loss function is presented in Figure 11.3.

The approach recommended by Taguchi for the quality of product design is a two-step process: a robust design technique, followed by tolerance design. The robust design technique attempts to reduce variation in a product. Instead of controlling the sources of variation, robust design attempts to reduce the design's sensitivity to those sources of variation. The focus of tolerance design, on the other hand, is to determine the extent of variation that is allowable in both the design and noise factors. The outcomes of this method are tolerances that minimize the manufacturing and lifetime costs associated with product. The rationale of this

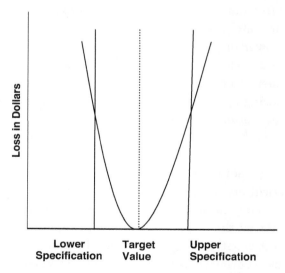

Figure 11.3 Taguchi's quality loss function

approach is to arrive at tolerance or specification limits as identified by parameter design, and not by convention.

The operational steps for determining a robust design require the use of the statistical methodology of experimental design, and involve the following steps[6]:

1. State the problem and the objectives.

   - It is important that problem definition and determination of objectives are the results of a team effort, where all interested parties are included.

2. Identify control parameters, and list responses and sources of noise.

   - Control parameters (also called design parameters) that identify the product features and their nominal or target settings can be specified by the product designers.
   - The outcome of this step is a parameter diagram (P-diagram), and is a prerequisite for every development project as it clearly defines the scope of development.

3. Design the experiment.
4. Run the experiment and predict improved parameter settings.
5. Run a confirmation experiment.
6. If the objective is not met, then go back to step 2 and repeat the process. Otherwise, the improved design can be adopted.

With the robust and improved parameter design in place, the next step is to allow deviations of the parameters from the nominal or target values. This requires a trade-off between the added costs of tighter tolerances and the benefits to the customer. The end result of this step is the tolerance design with lower and upper specification limits. The advantage of this approach of determining parameter design followed by tolerance design is that high-quality, successful products can be produced at low cost.

While a detailed discussion of Taguchi's loss function and his robust design methodology requires some expertise in the statistical methodology of design of experiments, the methodology is relevant and applicable to a project environment. The following are case study examples of the application of Taguchi's robust design methodology in a project environment.[7]

Example 1.   Robust Ground-to-Air Communication Receiver
A critical component of the ground-to-air communication receivers used in aircraft is the FM demodulator. The objective of the design of the demodulator is to minimize the bit error rate (BER). The key underlying concept of demodulation is the conversion of a received radio frequency signal to baseband signal, sampling it at the midpoints of the bits, and then, using a threshold voltage value, the bit needs to be identified as 0 or 1. Efforts expended by focusing on this main concept can save considerable design time and cost. The use of Taguchi's robust design methodology enabled designers to arrive at an optimum design for the FM demodulator that reduced the bit error rate (BER) by 37 percent.

Example 2.   Robust Paper Feeder Design
One of the problems that Kodak's copy machine manufacturing division had to contend with was improving the reliability of its paper feeder. The objective was to reduce the mean time between failures from 2500 sheets to 40,000 sheets. Instead of using the more expensive and time-consuming traditional method, which required feeding tens of thousands of sheets to determine the failure rate, the quality design team used the robust design methodology.

This approach provided a drastic improvement in the method of observing failures by focusing on the time needed for a sheet of paper to arrive at a sensor after the command was issued. Arrival time was affected by several noise factors, such as the weight of the paper, smoothness, and humidity—and if the paper hit the sensor outside of the design window, the result was feeding failure. To cause fewer failures, an improved design that reduced variation in arrival time was needed. The design team needed new instrumentation for measuring arrival time. Completion of this project using the traditional approach would have taken months. However, through the use of the robust design methodology, the design team completed the project in a fraction of that time and at a much lower cost.

These two short examples illustrate the superiority of Taguchi's robust design methodology when compared to traditional design methods. It has the clear advantage of reducing product development cycle time at much lower cost, while achieving the technological limits of the design concept. The end result is that the product has high reliability from its introduction, which, in turn, translates into the creation of value and higher profits.

11.2.2 Management Processes

The second determinant of overall project quality is the management processes used in the project. These processes typically include project start-up, coordination, control, and close out. To ensure the quality of management processes, standard procedures and guidelines must be followed. Furthermore, the standard procedures used by the project firm should be flexible enough so that they can be adapted to meet the requirements of individual projects. There should also be a mechanism through which management processes can be monitored, so that appropriate corrective actions can be taken if the processes used do not meet the required standards of quality. Finally, the personnel who are responsible for managing processes should be qualified and well trained.

11.2.3 Quality Planning

This phase of quality management is concerned with identifying the quality standards relevant to a particular project, and the associated planning necessary to meet those standards. Quality planning should be performed concurrently with other project planning processes during the development of the overall project management plan. This is because changes to the product that may be required to meet identified product standards can cause ripple effects with other aspects of the project management plan, such as cost and/or schedule adjustments. It also likely that the product quality desired as result of quality planning may necessitate a detailed risk analysis of an identified problem.[8]

The input, tools and techniques, and output of the quality planning process are presented in Figure 11.2. One of the key inputs is the project scope statement, which clearly identifies major deliverables, project objectives, performance requirements, technical issues, and any constraints such as costs, schedules, or resources. All of these elements can affect quality planning, and, therefore, the costs associated with project quality.

Most of the tools and techniques that are used for quality planning in the traditional manufacturing environment can also be used for projects. During the earlier discussion on Taguchi's robust design, we noted the use of the statistical method of design of experiments (DOE). Because this methodology is an extensively used quality planning technique, a brief overview appears below.

The statistical methodology of DOE is used to identify specific factors or variables that have an influence both in the development stages of a product or process, and during production. It helps to reduce the sensitivity of the performance of the product to sources of variation caused by noise factors, such as environmental or manufacturing differences.

The key advantage of using DOE is that it provides a statistical framework for simultaneously changing all of the important factors that have a direct influence on product performance, as opposed to making one change at a time. Through the analysis of experimental data, the methodology identifies the key factors that influence results, as well as the interactions and synergies among them, thereby facilitating the identification of the optimal conditions for product or process performance.[9] In the context of a project environment, DOE can be very useful, particularly during the design stages of a project's product. Readers interested in learning more about this methodology should refer to an advanced text in statistics.

Among the outputs from the quality planning process listed in Figure 11.2, the most important one is the quality management plan. It serves as an input to the overall project management plan, and provides the basis for other project quality-related issues, such as quality control, quality assurance, and continuous improvement.

11.2.4 Quality Assurance (QA)

Project **quality assurance** ensures that all processes involved in creating and delivering the product—such as design, engineering, implementation, and testing—are done right the first time. It also ensures that all project management processes are correctly followed from the outset, and that the best results in terms of project cost, time, and quality will be achieved. In addition, QA provides the impetus for continuous process improvement, an iterative framework for quality improvement at all levels that reduces waste and other non-value-adding activities.

11.2.5 Quality Control

Quality control is a process by which outputs are monitored to see if they conform to design specifications. However, unlike a typical manufacturing process where outputs can be repeatedly monitored, a project consists of only one product, with no margin for error. Consequently, the emphasis of quality control in a project is to ensure that output from every project process from the very outset is defect-free. In this way, the focus of quality is more on quality assurance—in other words, doing everything right the first time—rather than on quality control.

There are, however, several basic tools and techniques that can be used for quality control, including cause-and-effect diagrams, control charts, flowcharts, histograms, Pareto charts, run charts, and scatter diagrams. These tools, which are extensively used in traditional manufacturing environments, can also be used in a project context. Interested readers can find a wealth of information on these techniques in any statistical quality control text.

11.2.6 Corporate Culture

To achieve the highest level of quality in a project, a quality culture should pervade the entire project organization. In everything they do, every employee should focus on doing it right the first time, and making quality the primary emphasis of their job. To achieve this quality culture, the impetus should come from top management; in fact, the CEO of the project organization should assume the role of "Quality Champion." In addition, every employee should receive adequate training on quality methods, and should be given the responsibility for achieving and managing the quality of their tasks.

11.3 TOTAL QUALITY MANAGEMENT (TQM) IN PROJECTS

From a quality perspective, projects have very little margin of error. The entire thrust, in terms of quality, is doing everything right the first time. To achieve this goal, project organizations should embrace TQM. **Total quality management** is not a technique, but instead is a philosophy that involves the quest for quality by everyone in an organization. Total quality management is based on the following key principles, which should be adopted by any project organization that wishes to improve the quality of projects that it sponsors.

1. *Customer satisfaction*—In a project context, this involves satisfying the requirements and expectations of both external and internal customers (employees of the project organization).
2. *Continuous improvement*—This element of TQM emphasizes that improving the process of converting inputs to outputs in a project, as well as improvements to all associated procedures, is never ending. It covers all of the equipment used, procedures, methods, and employees of the project organization.
3. *Competitive benchmarking*—As a quality management tool, benchmarking involves learning from the practices, processes, and quality improvement methods of best-in-class competitors. To use the benchmarking tool, the project organization should first clearly understand and document the details of its own processes. The second step is to understand the processes and the best practices of companies that are known for their quality excellence, and to compare their performance with that of the project organization. From the lessons learned, the project organization should take necessary steps to close any performance gaps.
4. *Empowering employees*—This element of TQM advocates giving project organization employees the authority and responsibility for initiating and implementing quality improvement programs on their own. Such empowerment not only places the responsibility for achieving quality with those who are best qualified, but also serves as a strong motivating factor. However, continuous improvements in quality can be achieved only if *all* employees are provided with opportunities to enhance their skills and competencies. For this reason, employees must be educated and trained in quality improvement methods before they are empowered. In addition, they must be provided and familiarized with the proper tools, standards, policies, and procedures.
5. *Team orientation*—The use of teams and teamwork fosters cooperation and promotes synergy and a sense of shared values among the employees of the project organization.
6. *Decisions based on actual data*—The project organization collects and analyzes data, and decisions on quality improvements are based on the results of this data analysis. Consequently, decisions

made are based on pure facts, rather than opinions. A number of both quantitative and qualitative tools can be used for data analysis. Some of the quantitative tools used include statistical process control charts, cause-and-effect diagrams, and Pareto charts. In addition, some of the qualitative methods used include audits, reviews, and benchmarking.
7. *Supplier involvement*—Total quality management transcends organizational boundaries. If a project firm treats suppliers as long-term partners in the quality improvement process, its suppliers have a vital stake in providing quality raw materials and services. This kind of supplier involvement will also reduce the frequency and need to conduct quality inspections for inputs like raw materials and components.

11.4 QUALITY MANAGEMENT METHODS FOR A PROJECT ORGANIZATION

The following quality management methods are available for project organizations that have embraced the philosophy of TQM and continuous improvement.[10]

- *Certification*—Certification is a process by which an independent, neutral third party certifies that the product of a company and the processes used by it to produce that product meet acceptable standards. The International Standards Organization (ISO) is one such body. ISO is a network composed of National Standards Institutes from over 147 countries. Its purpose is to promote worldwide standards for improvements in quality, operating efficiency, productivity, and cost reduction. It works in close collaboration with many international organizations, governments, industries, and consumer representatives. The obvious advantage of ISO certification is that companies find it easier to conduct business worldwide.

  The ISO 10006:2003, titled "Quality Management Systems: Guidelines for Quality Management in Projects" is the new series of international standards for projects and programs. These standards provide structured, step-by-step guidelines for optimal management of all processes for any type of project. For project organizations without a quality management system in place, the ISO 10006 certification and registration process will be extremely beneficial.

In addition to ISO, other relevant certifying bodies for project organizations include Prince 2, MSP, IPMA, and PMI.

- *Accreditation*—Accreditation is an external evaluation of an organization's products, processes, and systems based on widely accepted standards and criteria, and is most frequently used in the education and healthcare sectors. Its main purpose is consumer protection. In a project management context, the degree and nondegree programs of the Project Management Institute (PMI) are accredited by the Global Accreditation Center for Project Management.
- *Quality award models*—In various countries, a number of quality awards are presented to companies that have achieved outstanding quality improvements. Some of these awards include the Malcolm Baldridge Award in the United States, the Deming Prize in Japan, the European Quality Award, and the International Project Management Award. These awards are based on quality and other established criteria in areas such as leadership, strategic planning, customer and market focus, information and analysis, human resource focus, and process management business results.

  Companies that win quality awards demonstrate excellence in each of these areas. For example, the International Project Management Award, which is based on the project excellence model developed by the German Project Management Association (www.gpm-ipma.de), uses nine assessment criteria divided into two categories and applies to any type of project. Project organizations that wish initiate quality improvement programs can use this model, which appears in Table 11.1.
- *Benchmarking*—We have already discussed benchmarking as a quality tool in the context of TQM. While a variety of benchmarking

Table 11.1 Assessment criteria for project excellence award (1000 Points)[11]

Project management (500 points)	Project results (500 points)
Project objectives (140 points)	Customer results (180 points)
Leadership (80 points)	People results (80 points)
People focus (70 points)	
Resources utilization (70 points)	Other stakeholders results (80 points)
Processes content and management (140 points)	Key performance and project results (180 points)

models are available, project management maturity models are the ones that are most frequently used in project organizations. More information about one of these models, the Berkeley Project Management Process Maturity Model, is available from Ibbs and Kwak.[12]

- *Audits and Reviews*—An audit is a process of systematic evaluation that can be used to ensure project quality. It is based on criteria that are composed of policies, procedures, and requirements. From the perspective of quality, an audit can uncover performance gaps that can lead to corrective action and subsequent quality improvements.

  For project organizations, the criteria used for an audit depends on the particular project management approach used. Among the variety of methods used in a project management audit are

  - *Document analysis*—involves analysis of documents such as the WBS, milestone charts, bar charts, project progress reports, etc., to determine the project management competence of the organization.
  - *Interviews*—these follow document analysis and are conducted to uncover detailed information. Interviews should be conducted not only with the members of the project team, but also with customers and suppliers to gather different perspectives of problems and performance in terms of quality.
  - *Observation*—auditors gather additional information by observing the day-to-day activities of the project organization.
  - *Self-assessment*—involves individual self-evaluation of project management competence by members of the project organization.

  From the point of view of quality management, the purpose of these project management audit methods is to identify shortcomings, so that corrective actions can be taken to improve quality. Reviews are similar to audits, but less formal.

11.4.1   The Six Sigma Methodology

The **Six Sigma methodology**, which helps to greatly reduce errors and defects in products, processes, and even services, originated as a process and quality improvement initiative at Motorola Corporation. Due to the unprecedented quality improvements that Motorola achieved, this methodology became very popular, and was adopted by other industries all over the world. At this level of quality performance, only 3.4 defects occur per million opportunities over the long term.

Six Sigma can be defined in a variety of ways. In a focused context, it involves the rigorous application of both statistical and nonstatistical methods to reduce the amount of variation in any given process, which implies greater efficiency and lower costs. The application of the methodology in a broader context can be described as a proactive management philosophy aimed at problem solving and performance improvement through the elimination of waste and non-value-adding activities throughout the organization. The expected benefits of its use in this context are reduced costs and increased customer satisfaction, as well as the creation of value.

11.4.2 The Six Sigma Model for Projects

Regardless of how Six Sigma is defined, it can be readily adopted to a project environment and applied to projects that require incremental improvements, as well as to those that require significant changes. The procedure comprises steps and is presented in Figure 11.4. A brief description of each of the steps presented in the Six Sigma model is given below.[13] In a project context, however, note that the specific details in each of these steps will vary from project to project.

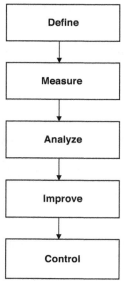

Figure 11.4 Six Sigma model for projects

- *Define*—Determine the quality characteristics of the output from the project's process that are critical to customer satisfaction. Identify any mismatch between these characteristics and the operating capabilities of the project's process. Document the existing process using a flow chart or process chart. If appropriate, redefine new performance objectives.
- *Measure*—Determine in quantifiable terms the work of the project's process. Determine what to measure, sources of data, and a data collection plan.
- *Analyze*—Use the appropriate quality planning or quality control tools and techniques to analyze the data collected. Determine what improvements are necessary, or whether a major redesign is required.
- *Improve*—Based on the new performance objectives defined, modify or redesign existing methods and implement the changes.
- *Control*—Monitor the project process to ensure that required performance levels are maintained.

11.4.3 Application of Six Sigma in Software Project Management

This example was adapted from "Reinforcing Six Sigma in Project Management," by Mahesh Chinnagiri.[14]

1. *Define*
   - Objective: Design a defect prevention system through the enhancement of intrinsic capabilities of various software processes.

2. *Measure*
   - Classify the various software projects into different groups.
   - Conduct a failure mode and effects analysis (FMEA) for each group of projects.
   - Identify the weak areas from the above FMEA analysis.

3. *Analysis*
   - Conduct a fault tree analysis (FTA) pertaining to the failure modes that have the greatest impact on the system as whole.
   - Analyze the occurrence paths, causes, and probabilities.
   - Determine the phases that require further analysis.

- Determine and establish optimal conditions and standards for various stages in the development process for each of the above identified phases to ensure defect prevention.
- Select a sample of completed projects from each group and analyze actual conditions.
- Compare the two previous steps and determine deviations.

4. *Improve*
   - Implement the necessary corrective measures or improvement for each deviation.
   - Using appropriate statistical tools such as quality function deployment (QFD), design of experiments, etc., institute the operating standards and preventive procedures to maintain the optimal performance condition.

5. *Control*
   - Validate the new system and monitor the system for improved performance.

11.5 QUALITY STANDARDS FOR PROJECTS

No quality improvement program will have any meaning without adequate standards for comparison. In a project environment, standards can be either generic (applicable to all types of projects) or specific (relevant to a specific type of project). In addition, other types of standards that may be relevant to quality management in project organizations include the following[15]:

- *Product standards*—Standards relating to the required functionality of the product and design.
- *Project standards*—Standard procedures relating to the content and management processes to be used; varies from project to project.
- *Project management standards*—Guidelines for the different management approaches that can be used to ensure project quality. Prince 2 and PMI's PMBOK are examples of generic project management standards.

- *Other standards*—In addition to the above standards, a project organization may have other standards for procurement, health and safety regulations, etc.

The issue of quality management in projects is one that is not sufficiently addressed in the project management literature. Indeed, many organizations have tended to view the drives for quality and the initiatives to improve project management performance as two different strategic thrusts rather than recognizing the natural complementarity that exists between these activities.[16] There is no question that managing projects for value must contain a strong emphasis on identifying and improving quality management, as quality has direct impact on our organizations' bottom lines and customer satisfaction. Put in this perspective, developing a systematic process for enhancing quality in our projects is simply a natural offshoot from any desire to enhance project value.

REFERENCES

1. (2008) *A Guide to the Project Management Body of Knowledge (PMBOK Guide)*, 4th ed., Newtown Square, PA: Project Management Institute.
2. PMBOK (2008) *Ibid.*
3. PMBOK (2008) *Ibid.*, p. 182.
4. PMBOK (2008) *Ibid.*, p. 183.
5. Turner, R. J. (2008) In R. Turner (Ed.), *The Handbook of Project-Based Management*, 2nd ed., Berkshire, UK: McGraw-Hill, p. 149.
6. Buyske, S., and Trout, R. (n.d.) Robust design and Taguchi methods, Lecture 10. www.stat.rutgers.edu/~buyski/591/lect10.pdf
7. Phadke, M. S. (n.d.) Robust design (Taguchi method) case studies. www.isixsigma.com/library/content/c020311d.asp
8. PMBOK (2008) *Ibid.*
9. PMBOK (2008) *Ibid.*
10. Huemann, M. (2004) Improving quality in projects and programs. In: P.W.G. Morris, and J. K., Pinto, (Eds.), *The Wiley Guide to Managing Projects*. Hoboken, NJ: Wiley, pp. 907–925.
11. Huemann, M. (2004) op cit., p. 928.
12. Ibbs, W., and Kwak, Y. (n.d.) (www.ce.berkeley.edu/pmroi/calculating-PMROI.pdf)

13. Krajewski, L. J., Ritzman, L. P., and Malhotra, M. K. (2009) *Operations Management Processes and Value Chains*, 9th ed., Upper Saddle River, NJ: Prentice-Hall.
14. Chinnagiri, M. (n.d.) Reinforcing Six Sigma in project management, www.qaiindia.com/Conferences/publishing/
15. Huemann, M. (2004) *Ibid.*
16. Kerzner, H. (2011) *Project Management Best Practices: Achieving Global Excellence.* 2nd ed., New York: Wiley.

KEY TERMS

Project quality
Quality engineering
Quality loss function
Robust design methodology

Quality assurance
Quality control
Total quality management (TQM)
Six sigma methodology

Chapter 12

Integrating Cost and Value in Projects

LEARNING OBJECTIVES

- Discover what is meant by "the value chain."
- Demonstrate how to perform a project value chain analysis.
- Examine the sources and strategies for integrating cost and value in projects.

It should be evident from the content of the preceding eleven chapters that achieving project success hinges on effectively managing both project costs and value. The discussions in these chapters, however, focused on the various issues and approaches to *individually* managing these two critical project success factors. In this chapter, our goal is to unify the concepts and processes of project cost and value management so that these two vital decision areas can be treated as an integrated whole.

We begin our discussion with the concept of a **project value chain**, which demonstrates the inseparable and interwoven nature of project costs and value. In subsequent sections of this chapter, we present various approaches to simultaneously managing cost and value to achieve project success.

12.1 THE PROJECT VALUE CHAIN

The concept of "value chain" was first articulated by Michael Porter.[1] The essence of Porter's value chain concept is that an organization needs to manage a set of basic activities to create value that in monetary terms is greater than the cost of providing the organization's product or service. The end result of effectively managing these activities is that

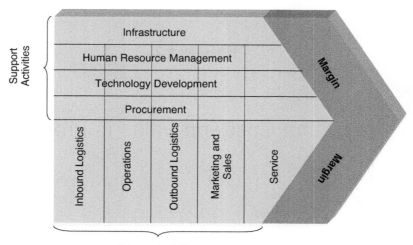

Figure 12.1   Porter's value chain model

the organization generates a profit margin and sustains its competitive advantage in the marketplace. Figure 12.1 presents the basic model of Porter's value chain.

In Porter's value chain model, there are two distinct sets of activities: primary activities and support activities. In a typical organization, primary activities are directly responsible for creating or delivering the product or service. They comprise inbound logistics, operations, outbound logistics, marketing and sales, and service. The support activities that are linked to each of these primary activities help to improve their effectiveness or efficiency. Support activities can be grouped into procurement, technology development (including R&D), human resource management, and infrastructure (systems for planning, finance, quality, information management, etc.).

Value chain analysis evaluates the value that each particular activity adds to the organization's products or services. The goal is to make it possible to produce a product or service that delivers the value customers want, and for which they are willing to pay a price. Porter contends that the ability to perform these specific activities and to manage the linkages between them is a source of value and competitive advantage to the firm.[2]

Project value chain analysis provides an entirely different perspective on what constitutes project success. While the traditional measures of project success, such as schedule, cost, and technical validity, are still relevant, project value is much more than these internal measures,

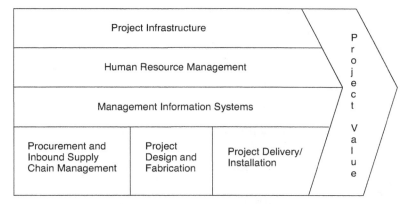

Figure 12.2 Project value chain[3]

and ultimately it is the customer who determines that value. In project management, the adoption of the value chain concept offers considerable potential for ensuring that projects meet the unique needs of customers.

A basic model of the project value chain is presented in Figure 12.2. It is similar to Porter's value chain model in that the project value chain is also composed of primary and secondary activities. The primary activities are procurement and inbound supply chain management, project design or fabrication, and project delivery or installation. All of these activities are directly responsible for creating the project's facility, and therefore, project value. Secondary activities include the project infrastructure, management information, and human resource management systems.

The critical elements of the project infrastructure that enhance or optimize project value are cost management, scope management, schedule management, quality management, configuration management, and change control systems. Within the project infrastructure, cost management is an umbrella system that is concerned with all cost-related subsystems such as accounting and finance, and all cost-related activities such as project cost estimation, budgeting, and cost control.

12.2 PROJECT VALUE CHAIN ANALYSIS

The primary thrust of project value chain analysis is to help companies understand, without any ambiguity, how they can create and deliver the value that is most critical to the customer and their activities. A project organization that can successfully accomplish this goal has a clear competitive advantage. An important prerequisite of a successful value chain analysis is that the company has thorough knowledge of not only its own strengths and weaknesses, but also those of the potential customer. Armed with this knowledge, the project organization is better equipped to address the unique and critical needs of the customer's business operations.

Performing a successful value chain analysis requires the following four steps:[4]

Step 1: Construct a value chain that is unique to the customer—Because each project is unique, the first step in developing a value chain is for the project organization to clearly understand its role in the customer's value chain. To review, a project value chain consists of a sequential chain of activities that begins with acquisition of raw materials from suppliers, progresses through the development of the project, and concludes with the delivery of the completed project to the customer. While each activity in the chain has the potential to add value, the key role of the project organization in the value-adding process is to provide superior quality at lower costs and risks by effectively designing and fabricating the project facility.

To develop a meaningful value chain strategy, the project organization should be fully aware of the areas in which they can make the most significant contributions that will benefit the customer. This requires that the project firm clearly understand its own competitive strengths—the activities that the organization does well, and that can add value. For example, one project organization may use superior technology and design capabilities to add value to client operations, while another may enhance value by using global links to suppliers and distribution capabilities to acquire high-quality raw materials at a lower cost.

Conversely, the project organization also needs to understand its weaknesses—those activities that it does not do well relative to the competition. The key is to recognize areas where the project organization can best use it competitive strengths, instead of attempting every aspect of the client's operations.

Step 2: Identify the client's value drivers—At this step, value analysis should focus on identifying the activities and linkages of client operations that distinctly add value and are sources of competitive advantage for the client. For example, the focus should be on activities that enable the client to achieve a cost advantage or produce a distinctly unique product, which are value drivers that create differentiation. By focusing on the client's value drivers, the project organization can develop project design choices that can provide the customer with maximum competitive benefits.

Step 3: Clearly identify the weaknesses of the project organization—To develop and deliver a successful project, it is necessary that the project organization be superior in every aspect of the project. However, the project organization cannot provide value enhancement in every aspect of the client's project—there will be areas in which the project firm is clearly not superior to the competition. The key is to recognize those activities in which the firm is inferior and operates at a competitive disadvantage, and to outsource them to those who can do them well. As long as the project organization can provide the client with a source of competitive advantage by engaging in activities where it can enhance value, while at the same time eliminating the sources of competitive disadvantage, the end result will be a successful project that can reap superior returns both for the client and the project firm.

Step 4: Target the activities and linkages that have the greatest impact on project value—In the final step of the project value chain analysis, the project organization and the client should come to clear understanding of the options and alternatives for activities where maximum value enhancement can be achieved, and for activities that require improvement. This will ensure that sources of competitive disadvantages can be eliminated. It is in this final step of value chain analysis that the project organization can set the stage for achieving project excellence. To do so requires two things: first, the project firm must view project success from the client's perspective, instead of through the usual parameters of time, budget, and specifications. Second, the project firm must develop a partnership with the client so that it can play an active role in the success of the overall project, rather than playing the passive role of a contractor fulfilling a contractual obligation.

In this framework of a project value chain, subsequent sections of this chapter explore various issues and strategies for integrating cost and value in projects.

12.3 SOURCES AND STRATEGIES FOR INTEGRATING COST AND VALUE IN PROJECTS

There are several fertile areas that provide significant opportunities for enhancing project value while minimizing costs. While these were addressed in previous chapters, we are revisiting them to provide the necessary platform for developing appropriate strategies for project cost and value integration.

12.3.1 The Project's Inbound Supply Chain

The inbound portion of the project supply chain that involves procurement of raw materials is extremely complex, with many tiers of suppliers. For many project organizations, materials and components purchased from outside vendors represent a large portion (as high as 60 percent) of the total cost of the project. Employing effective procurement strategies that focus on acquiring quality raw materials at a reasonable cost can significantly reduce the overall project cost while simultaneously adding value to the project value chain.

Strategies such as Supplier Kaizen, international sourcing, partnering based on trust, vendor development, information and risk sharing, long-term strategic alliances with suppliers, and integrating quality management activities of suppliers through total quality management (TQM) will significantly reduce procurement and inventory costs, shorten lead times, and improve the quality of purchased materials—all of which improve the value of the total project. In addition, the use of lean manufacturing techniques to achieve tighter integration with suppliers and just-in-time delivery of parts can facilitate reduction in raw material inventory costs and improved efficiency, thereby enhancing project value. These supply chain management strategies were discussed in detail in Chapter 10.

12.3.2 Project Design

The first attempt in creating project value originates at the project design stage, where the degree of success and the challenges in integrating cost and value depends, to a large extent, on the project's complexity. The more complex the project, the more challenging is the task of integration. However, once integration is achieved, the reward to all project stakeholders is enormous. Regardless of the nature of the project, however, several strategies that have proven to be successful in the traditional manufacturing environment can be employed to enhance value and reduce cost. These appear in Table 12.1.

During the project design stage, performing value engineering can greatly facilitate project cost and value integration. Before delving any deeper into the use of this approach, however, a clear distinction needs to be made between value engineering and design engineering.

Table 12.1 Strategies for integrating cost and value

Method	Advantages	Examples
1. Inbound supply chain management	Primary means to minimize costs through vendor management	Kaizen, international sourcing, reverse auctions, project partnering
2. Project design	Allows integration of cost and value through enhanced design, creative problem solving, and collaboration	Value engineering, concurrent engineering, Kano modeling
3. New product development	Links procurement and design to new product development, allows for gated reviews, rapid product modification response	Total quality management (TQM), Six Sigma, lean manufacturing, target costing
4. Project delivery management	Enhances value to customer through carefully managed delivery cycle	Life cycle costing, turnkey project management

When design engineers are presented with a project concept, they use their decision process to come to an understanding of the product that they think the customer wants. Based on this understanding, and using engineering principles, they generate a design for the product that they think is optimum for the customer. The decision process of design engineers, however, seldom utilizes functions, function logic diagrams, or value-based comparative analysis methods—even though any key parameter missed in the design may greatly affect customer satisfaction with the product. Also, design engineers focus primarily on efficient designs to keep project costs down. Consequently, these engineers may find an alternative design unacceptable, even though it may have the greatest potential to increase value.

The value engineering method, on the other hand, uses a value-based decision-making approach to ensure that resources such as time, money, and personnel expertise are directed toward the solutions that have the highest potential for meeting customer needs at optimum cost.

Furthermore, this method seeks to generate the largest number of creative solutions to broaden the potential for achieving better value.

For this reason, value engineering should be used in tandem with design engineering, because it provides appropriate checks and balances to the existing design process. In addition, at the completion of the design process, an approach that uses a combination of design engineering and value engineering analyses leads to project design that ultimately has the potential to integrate both cost and value while generating the optimum product that the customer wants.[5]

Concurrent engineering is another potentially useful strategy for integrating cost and value in the project design phase. In a project context, concurrent engineering means bringing project design and development personnel together very early in the design process to simultaneously develop the design for the project and the process for completing the project. This concept can be further extended to create a cross-functional team that includes material specialists, marketing and purchasing personnel, suppliers, and even the customer. The objective, of course, is to come up with a project design that not only incorporates what the customer wants, but also factors into account the capabilities and constraints of the project organization and its suppliers.

There are several key advantages to this approach:

- The project team is able to identify, very early in the project lifecycle, resource availability and capabilities.
- The impact of the design on cost and quality aspects of the project can be understood very early, and potential conflicts can be greatly reduced further downstream, during the project development process.
- Aspects of design or procurement of critical tooling that may entail long lead times can also be identified early. This can have a very favorable impact on both the project schedule and cost.
- Early recognition of the technical feasibility of the design can prevent future problems during project development.

Clearly, the application of concurrent engineering during project design has considerable potential for defining and creating project value, while simultaneously providing opportunities for minimizing project life-cycle costs. In fact, the application of concurrent engineering can be carried out by the value engineering team that is responsible for integrating and optimizing project cost and value.

Sources and Strategies for Integrating Cost and Value in Projects

Finally, the **Kano model** of customer satisfaction shows great promise for integrating cost and value during project design.[6] This model presents a unique way for classifying project design characteristics, based on how the customer perceives these characteristics and their effect on customer satisfaction. The Kano model, which appears in Figure 12.3, depicts the relationship between customer product needs and customer satisfaction.

In the Kano model, product design characteristics are classified into three categories: threshold attributes, performance attributes, and excitement attributes. In the context of a project environment, the **threshold (or "must-be") attributes** are the basic "must have" characteristics in the design of the project. Any absence or deficiency in these characteristics will result in extreme customer dissatisfaction and diminished project value. For example, in building a house, a client has the right to expect the roof to keep out the rain, the heat to work, walls to remain upright, and so forth. Any enhancements to this basic functionality (frills) will only increase costs without any appreciable increase in customer satisfaction or project value, because the customer already expects them to be present.

Performance ("one-dimensional") attributes are those that the customer expects. Their presence results in customer satisfaction and their absence or deficiency will result in extreme customer dissatisfaction. These attributes are considered to have a straight, linear relationship between requirements and satisfaction. An example of a performance attribute is gas mileage on a new car. The better the mileage, the greater

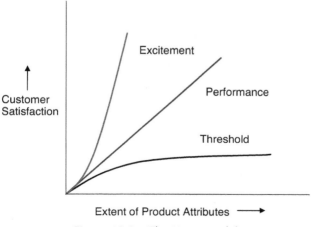

Figure 12.3 The Kano model

the customer satisfaction; the worse the mileage, the less happy the customer will be with the car. The price that the customer is willing to pay for the project and the value delivered by the project are determined by the performance attributes.

Finally, **excitement attributes** often satisfy latent needs that the customer is not yet aware of, and often have the greatest effect on customer satisfaction. Because the customer may not expect them, including these characteristics in a product is likely to have a powerful positive impact on satisfaction. For example, a state-of-the-art sound system or GPS routing system may not be something the customer expects to find in the car she is considering purchasing, but both have the potential to profoundly affect her satisfaction with the product.[7]

A relatively simple approach that the project design team can adopt to implement the Kano model analysis is to ask the customer the two questions in Table 12.2 for each project design characteristic ("Kano model analysis," www.ucalgary.ca).[8]

In general, for threshold ("must-be") attributes, customers will likely have a "neutral" response to the first question. However, the response will be "dissatisfied" for the second question, because exclusion of these characteristics will severely impair the project's functionality. Customer responses to these questions on performance or excitement attributes will lead to the inclusion or exclusion of project design characteristics. This information is ultimately used as a basis for deciding on whether sufficient value can be added to the product for the cost of including the feature.

Ultimately, the decision to include or exclude these attributes should be based on a well thought out cost–benefit analysis. In fact, for each

Table 12.2 Customer satisfaction ratings of design attributes for Kano model analysis

	1. Satisfied	2. Neutral (OK)	3. Dissatisfied	4. Indifferent
1. How would you rate your satisfaction if the product has this attribute?		√		
2. How would you rate your satisfaction if the product does NOT have this attribute?			√	

characteristic the customer rates as important, asking the question, "How much additional cost are you willing to incur for this characteristic or more of this characteristic?" will help in making trade-off decisions. This can be especially important for performance attributes. Note that characteristics that receive "indifferent" responses from the customer should not be ignored outright, as they may be critical to the product's functionality, or may be required for reasons other than customer satisfaction.

In a project environment, the Kano model is an extremely useful methodology for identifying customer needs, determining functional requirements, and concept development. When used in conjunction with value analysis, the model can play a vital role in integrating project cost and value.

12.3.3 Project Development

As we noted in Chapter 10, project development is where the project's product is actually created.[9] The extent to which project value is created and enhanced in this phase depends on the efficiency and effectiveness of the previous project phases—the procurement and project design phases. The extent of complexities and challenges encountered in the project development phase is largely determined by whether the project is relatively routine or highly complex. Regardless of complexity, using strategies like lean manufacturing techniques and applying stringent quality control procedures like Six Sigma and TQM can add value and reduce project costs by eliminating waste and unnecessary inventories, by improving process yields and process and product quality, and by shortening process lead times.

The application of lean manufacturing during the project development stage has the greatest potential to reduce project cost while maintaining desired project value. Lean manufacturing, however, is not a short-term process. Like TQM, it is a management philosophy and is a continuous improvement process that requires employee training, employee involvement, and employee empowerment. Lean manufacturing frequently utilizes a combination of techniques such as total cost management, concurrent engineering, team-based work arrangements, supportive supply chain relations, and integrated product development. The end results are increased process yields, reduction of waste in the form of reduced scrap, shorter process queues and smaller inventory, and elimination of costly waste streams.[10]

With lean manufacturing, cost reduction and value optimization in the project development stage are facilitated by effective cost management that requires the use of three primary tools: interorganizational cost management, value engineering and value analysis, and target costing.

Interorganizational cost management requires total integration of the supply chain, which was discussed in Chapter 10. The key issue lies in adopting a "target costing" philosophy across the supply chain established to support the project. **Value engineering and analysis**, as discussed earlier in this chapter, offer a primary means for engineering, or creating value through product design and post-project analysis for enhancing future projects as they are developed.

Target costing, in a project context, can be defined as a systematic approach to planning the cost of the project at a specified functionality and quality, to generate a certain level of profit at the project's anticipated selling price. Target cost is the total cost of the project, which includes direct costs, indirect costs, and life-cycle costs. Cost reduction approaches in target costing generally focus on reducing direct costs, because most cost management systems do not have an accurate way of determining indirect costs. Target costing should not be viewed simply as a tool for cost reduction and cost management, but as part of an overall profit management process. In fact, target costing should be an integral part of any project that involves the design of a new product. The specific steps in the target costing process are illustrated in Figure 12.4.

It can be seen from the figure that if target cost is greater than the estimated project cost, then no further analysis needed. On the other hand, if the target cost is less than the estimated cost of the project, a decision needs to be made as to whether or not to proceed with the project. The decision to continue with project development is appropriate if the options presented in Figure 12.4 can be implemented, with reducing the cost estimate as the most feasible alternative. This is where value engineering and value analysis can play vital role in the reduction of material procurement, engineering, and the manufacturing/development/ construction costs of the project. The least desirable option is to scrap the project, but this may be the only recourse left to avoid any further losses.

Sources and Strategies for Integrating Cost and Value in Projects

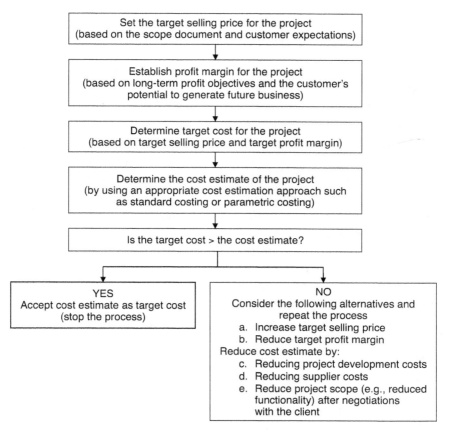

Figure 12.4 Target costing process for a project[11]

12.3.4 Project Delivery/Implementation

Project delivery offers its own set of challenges, depending on the nature of the implementation. Is it for an internal client? An external customer? To take advantage of a commercial opportunity? In cases, for example, where the customer is an external client, say, in the case of a large construction project, the actual project transfer represents a set of challenges and management issues that must be clearly understood. These challenges add an important dimension to the traditional, execution-based project life cycle that was thought to rapidly "ramp down" once the project had been developed. In other words, the tacit implication was that the project's challenges were generally mastered by the time of its termination, including transfer to the client. We know, in fact, that the delivery stage is in itself critical and a phase that must be carefully managed for project success.

Turnkey project management is an example of this perspective. **Turnkey projects** are defined as those in which the contractor provides a completed facility that includes all items necessary for use and occupancy. The result is an expansion of the project development life cycle, a phenomenon that Peter Morris has termed, "the management of projects," in recognition that it replaces the traditional, execution-based perspective on managing projects for immediate development and transfer.[12]

While this enhanced involvement in development and transfer increases the risk to the project organization, from a project value chain perspective, it also has the potential to enhance the value delivered by the project. At the same time, it presents the project organization with an entirely new set of challenges in that it is no longer enough to hand the completed project over to the satisfaction of the customer. Instead, the project needs to continue to deliver value to the customer throughout its life. This creates some serious cost implications for the project organization, which now has to contend with project costs not only from concept development to implementation, but for the entire project life cycle. This situation underscores the importance of decisions made during the early stages of the project, where they can have a significant impact on the overall value delivered by the project, as well as its life-cycle costs. An important tool for dealing with the cost and value implications in this "new world" of project delivery and implementation is project life-cycle costing analysis.

Life-cycle Costing

Life-cycle costing (LCC) is "a general method of economic evaluation which takes into account all relevant costs of a building design, system, component, material, practice or project over a given period of time, and adjusting for differences in the timing of those costs."[13] Life-cycle costing models can be used to track the costs of development, design, manufacturing, operations, maintenance, and disposal of a system or project facility over its useful life. These models relate cost components as a function of some independent or explanatory decision variables.

This relationship, known as a **cost estimating relationship** (CER), can be used to analyze the effect of changing the decision variables on one or more of these cost components, such as analyzing the effect of work design on labor costs. For example, in a project environment, we are aware that the slope of the learning curve depends on the type of manufacturing technology employed. A CER can help design engineers choose the most appropriate technology, as illustrated below:

Assume that Technology X incurs a lower labor cost to produce the first unit, but has a slower learning rate than Technology Y. All other things being equal, the choice of technology depends on the number of units to be produced and the cost of capital. If a smaller number of units are to be produced, then Technology X is the preferred choice as the labor costs are lower in the earlier stages of the corresponding learning curve. Also, if the cost of capital is high, Technology X might be preferred, since it displaces a substantial portion of the labor cost into the future. On the other hand, if a large number of units need to be produced, then employing Technology Y makes better sense.

What this example illustrates is that the importance of LCC models increases when the proportion of manufacturing, operations, and maintenance costs is greater than the proportion of design and development costs over the lifetime of the project.

Life-cycle costing models are particularly useful in the early stages of a project life cycle, when a number of alternatives exist, and the selection of an alternative has a noticeable influence on the total life-cycle cost. At the beginning stages of a project, LCC models provide a mechanism for evaluating alternative designs. Later, as work progresses on the project, these models can be used to evaluate proposed engineering changes. Through the diligent and proper use of LCC models, project engineers and managers can choose alternatives so that the life-cycle costs of the project are minimized while the required project value is maintained.

Life-cycle costing models are subject to the greatest degree of uncertainty at the outset of the project, during the conceptual design phase. This is because very little is known about the project, the activities required to design and develop the project, and the maintenance philosophy to be employed. This uncertainty declines as the project evolves and additional information becomes available. The two major uncertainties that LCC model developers encounter are the following:

Type 1: Uncertainty regarding the cost-generating activities during the project life cycle

Type 2: Uncertainty regarding the expected cost of these activities

Type 1 uncertainty is present in the case of a brand new project, for which few or no historical data exist. An example of this is the U.S. Space Station, where maintenance practices can be finalized only after sufficient operational experience is accumulated. This has led to a high level of uncertainty with respect to maintenance requirements for the equipment, as well as procedures for operating and maintaining launch vehicles and supporting facilities. As a result, the reliability and dependability of these systems are being carefully studied to determine the required frequency of scheduled maintenance.

The accuracy of LCC models where this type of uncertainty is present is relatively low, implying that their benefits may be somewhat limited to providing a framework for enumerating all possible cost drivers and promoting consistent data collection efforts throughout the life of the system. But even if this were the only use of the model, benefits would accrue from the available data when the time came to upgrade or build a second-generation system.

The second type of uncertainty (Type 2) is common to all LCC models; examples include future inflation rates, expected efficiency and utilization of resources, and failure rate of system components.

Analysts building LCC models should always trade off the desired level of accuracy with the cost of achieving that level. In the case of engineering projects that typically deal with improvement in existing systems, or the development of new generations of existing systems, it is possible to increase the accuracy of cost estimates by investing more effort in collecting and analyzing the underlying data.

Finally, LCC models should include all significant costs that are likely to arise over the project life cycle. This makes it imperative to develop an appropriate scheme to classify and structure cost components, based on the logical design of the project, familiar management concerns, and supporting data requirements. While a number of classification systems are available— including the cost accounting classification system, cash flow classification system, and work breakdown structure classification system—we will discuss a cost classification system that is based on the five phases of the project life cycle. In general, however, the particular classification system to be used depends on nature of the project, and the most appropriate one for a given project may combine the best features of several.

12.3.5 Costs of Project Life Cycle Employing the LCC Model

1. *Cost of the conceptual design phase*—This category highlights the costs associated with early efforts in the project life cycle, including feasibility studies, configuration analysis and selection, systems engineering, initial logistic analysis, and initial design. The costs in this phase usually increase with the degree of innovation involved, and are especially expensive for projects aimed at developing new technologies. Examples of such projects would include development of a new drug for AIDS, or the development of a permanently manned space station.
2. *Cost of advanced development and detailed design phases*—In this phase, the cost of planning and detailed design is presented, including the product and process design, preparation of final performance requirements, preparation of the work breakdown structure, schedule, budget and resource management plans, and the definition of procedures and management tools to be used throughout the life cycle of the project.
3. *Cost of the production phase*—The costs in this phase include costs associated with constructing new facilities or remodeling existing facilities for assembly, testing, production, and repair. Other costs included in this phase are the actual costs of equipment, labor, and material required for operations, as well as blueprint reproduction costs for engineering drawings and the costs associated with documenting production, assembly, and testing procedures.
4. *Cost of operating and maintaining the project facility*—This category identifies the costs surrounding the activities performed during the operational life of the project. These include the cost of personnel required for operations and maintenance, together with the cost of energy, spare parts, facilities, transportation, and inventory management. Design changes and system upgrade costs also fall into this category.
5. *Cost of divestment phase*—This category identifies costs that would be incurred when the useful life of the project's facility has expired and must be phased out. Parts and subassemblies must be inventoried,

sold for scrap or discarded, and, in some cases, it is necessary to take the system apart and dispose of its components safely.

The relative magnitude and timing of these cost components will vary from project to project. For example, in the case of research and development projects, the conceptual and advanced design phases may account for 50 percent of the life-cycle costs, where as in other routine projects these two phases may account for only 30 percent of the life-cycle costs.[14] Unfortunately, space limitations prevent us from presenting a detailed discussion of the actual steps involved in developing a LCC model. Interested readers, however, can use the references to learn more about it.

In summary, the life-cycle costing approach has great promise in integrating cost with project value as it looks at the entire project value chain in minimizing the total life-cycle costs of the project, while maintaining its value to the customer.

12.4 INTEGRATED VALUE AND RISK MANAGEMENT

In Chapter 8, we discussed in some detail the relationship between project value and project risk. Here, we reexamine this relationship from the perspective of project value and cost integration.

There are two important requirements for project success: to deliver value by way of meeting the needs of clients at the price they are willing to pay, and to minimize the impact of inevitable risks that can adversely affect project outcomes. The first requirement can be met by value management, which provides an effective way to maximize project value as defined by client requirements. The second requirement for project success is fulfilled by risk management, which provides an effective process for managing project risks.

Combining these two separated management processes into a single, formal integrated process is an excellent strategy that can maximize project value and return on investment while reducing uncertainty. As an integrated whole, these two processes provide the following conditions that are conducive for project success[15]:

1. *Clarity of purpose*—Value management provides the mechanism to clearly identify long-term project objectives in terms of benefits to the client. Risk management paves the way for the project team to minimize the uncertainty in delivering these benefits.

2. *Leadership and culture*—The two processes together promote leadership by providing a well-defined decision-making structure to maximize value and minimize risk. They also promote a culture for all project stakeholders to work together for a common goal.
3. *Communication*—Because both value and risk management require extensive consultation, effective communication channels are established that further improve decision making and problem resolution.
4. *Realistic and affordable budgets*—Clear identification of project objectives with full understanding of project risks facilitates establishment of realistic and achievable budgets. This, in turn, enables the value and risk management processes to proactively manage the project supply chain to deliver the required quality of materials and the final project facility within budget.
5. *Effective procurement*—The integrated process provides a project risk profile that enables the allocation of risk to those parties in the project supply chain who are best equipped to manage them, and provides the basis for selecting the most effective procurement strategy.
6. *Meeting project completion deadlines*—Adoption and application of the integrated value and risk management process from the very outset of the project provides adequate time to complete project design and preparation activities before a commitment is made to the construction of the project facility. This enhances the ability of the project team to accurately predict completion dates.
7. *Efficient project delivery*—The integrated process provides the structure for optimum delivery processes that lead to performance improvements, in terms of reduction in abortive time and waste during project design and construction activities. In this way, the integrated process not only provides a clear articulation of project objectives, but also maximizes efficiency in terms of cost, time, and quality.

Case Study: New Product Development at PING Golf, Inc.
One of the top performers in the golf equipment industry, PING Golf has long been known for their innovative designs and attention to quality. The golf equipment industry is generally defined as including (1) hard goods—golf clubs, including irons, wedges, putters, and fairway metals and drivers; (2) soft goods—clothing, bags, towels and ancillary equipment; and (3) golf balls. PING, headquartered in Phoenix, Arizona, is a privately-held company

that competes in the hard and soft goods categories and generally ranks in the top five manufacturers annually for revenues.

In recent years PING has sought to find new methods for improving new product development, including speed to market for new products, while minimizing the risk of product failures. The golf equipment industry thrives on new products and most golf club life cycles are generally thought to last no more than three years. Thus, a steady supply of new product designs and introductions is critical for continued success in this highly competitive market.

PING is one of the first golf equipment manufacturers to streamline its development cycle through the use of *finite element analysis* and design modeling, using sophisticated computer technologies. Traditionally, all new product ideas were first designed and then prototyped, after which they went through extensive and rigorous testing to determine if the design was viable, if it demonstrated strong performance characteristics, and if it could be developed for a reasonable cost. This development cycle was lengthy and made it difficult to bring new products to market at a reasonable rate. Finite element analysis, on the other hand, employs a supercomputer and on-screen design and simulated testing. In short, all new designs can be tested through computer simulations to determine performance characteristics before they are ever developed as prototypes, saving PING money and, more critically, time to market. John K. Solheim, Vice President of Engineering, refers to this process succinctly as, "Test, *then* Design."

PING has successfully integrated project value and risk management through streamlining its development process, eliminating unnecessary steps and bottlenecks, as it created a new product development process that is both efficient and cost-effective. The results are clearly evident in PING's G5 series of irons and fairway metals being rated the #1 selling irons and driver for 2005 and 2006.

Referring to this "Test, then Design" approach, Solheim noted: "We want to test the 'physics' of a principle and understand the variables and the relationship between those variables. Once we understand this 'knowledge landscape,' then we can design with confidence. The reverse of this is to pick a design geometry and test it to see if it works. In our paradigm of testing the physics first and designing within the parameters we know work, our designs are much more reliable in terms of quality and performance."

12.5 THE PROJECT COST AND VALUE INTEGRATION PROCESS

There are some essential prerequisites to the process of integrating project cost and value. First, it requires a team approach, with representation from all project stakeholders, including the client, project team, designers, people responsible for implementing the project facility, and all potential users. In essence, the team should be composed of all stakeholders who are knowledgeable about the project and have decision-making authority. It would helpful at this stage to bring the team together for "creative" workshops to generate ideas for integration. Second, the team should be willing to think "out of the box," utilize imagination, and forego the penchant to exercise pet solutions. Third, the team should focus on overall project objectives, targeting only those areas that are likely to yield maximum benefits. Finally, the team should be led by an experienced, cost-oriented project management facilitator.

The objective of project cost and value integration is to obtain, for each activity selected for analysis, the optimum value for every dollar spent. The process involves a sequence of five steps.[16]

1. *Generate information*—The objective of this step is to come up with an information base, or "cost model," and to select areas or activities in the project for detailed study. For each project activity, information should be generated by asking the following questions:

    - What is the exact nature of this activity?
    - What is the activity intended to accomplish?
    - What is the cost of performing this activity?
    - What is the potential value generated by this activity? (i.e., is this activity necessary?)
    - Is there another alternative than to do the same job, and at what cost?

    Some of the techniques that can be useful in answering these questions are functional analysis and cost–benefit analysis.

2. *Generate alternatives*—The objective of this step is to generate creative alternative solutions. For each project activity, think about alternative ways of meeting the same requirements or performing the same functions. Brainstorming sessions by the project cost and value integration team can generate many new ideas.

3. *Perform analysis*—The objective of this step is to evaluate each of the alternatives generated in the previous step, culminating in the selection of the one that presents the best combination of cost savings and required value. The appropriate questions to ask at this stage are

- What is the cost of each alternative?
- Which of the alternatives will meet essential functionality requirements?
- Which will meet essential functions and still offer the greatest cost savings?

One of the most useful techniques that can be employed at this stage is the life-cycle costing approach discussed earlier in this chapter.

4. *Generate a proposal*—The objective of this phase is to present the best sets of alternatives for achieving cost and value integration to appropriate decision makers. An important caveat in this step is to create a proposal that is most convincing, and to present it in the most effective way possible. A combination of techniques, such as narrative reports, schematic displays, graphic illustrations, and video clips of similar installations can be used to accomplish this.
5. *Issue a final report*—Finally, at the last stage it is important to define and, if possible, quantify the results in terms of which recommendations were actually approved and implemented. Techniques such as standardized comparative reporting can be useful in generating the final report.

In the final analysis, integrating cost and value in a project cannot be viewed as a short-term, quick-fix solution. It is an ongoing improvement process that requires the total commitment of all project stakeholders, and, in particular, the commitment of the project organization's top management. To accrue long-term benefits, the process should be started very early in the project and should continue throughout its life cycle.

In other words, as we have attempted to show, cost and value management do not begin with the commencement of the actual project planning and execution. Rather, the seeds for successful cost and value management are sown much earlier in the project's life, during procurement and supply chain development, and they run well past the point of project completion. In fact, cost and value management involve a number of critical delivery activities as the project is transferred to its intended users. Thus, in understanding project cost and value management, it is first important to adjust our thinking toward Morris's "management of projects" paradigm.

Likewise, it is critically important to understand that successfully managing for value requires us to embrace a systems thinking perspective. Certainly, we can find ways to occasionally "save money" on our projects, or find features that add value as we move through a development cycle. However, the theme of this book is to avoid a disconnected, fragmented mindset that treats cost and value management as issues to be addressed when it's convenient to do so.

The key is to understand that a model for cost and value management in projects involves a systematic, life-cycle philosophy based on their integration and used as an ongoing feature. This book has been our attempt to offer just such a framework for systematically integrating these two critical elements in successful projects: cost management and creation of value. An unyielding commitment to this process is vital for developing the kinds of projects that will provide our firms with a decided competitive advantage.

REFERENCES

1. Porter, M. (1985) *Competitive Advantage: Creating and Sustaining Superior Performance*. New York: Free Press.
2. Recklies, D. (n.d.) The value chain. www.themanager.org/models/ValueChain.htm
3. Pinto, J. K., Rouhiainen, P., Trailer, J. W. (1998) Customer-based project success: exploring a key to gaining competitive advantage in project organizations. *Project Management*, 4 (1), 6–12.
4. Pinto, J. K., and Rouhiainen, P. J. (2001) *Building Customer-based Project Organizations*. New York: Wiley.
5. Keenan, M., and Martin, S. (1997) But we already do it, and other misunderstandings. http://www.value-engineering.com/doitpapr.htm
6. (n.d.) Kano model analysis. www.ucalgary.ca
7. Sireli, Y., Kauffmann, P., and Ozan, E. (2007) Integration of Kano's model into QFD for multiple product design. *IEEE Transactions on Engineering Management*, 54 (2), 380–390.
8. (n.d.) Kano model analysis. *Ibid*.
9. Pinto, J. K. and Rouhiainen, P. (2001) *Ibid*.
10. (n.d.) http://www.iprod.auc.dk/misg/papers/creese.pdf
11. Cost management in lean manufacturing enterprises and the effects upon small and medium enterprises. (http://www.iprod.auc.dk/misg/papers/creese.pdf); Cooper, R., and Slagmulder, R. (1997) *Target Costing and Value Engineering*. Portland, OR: Productivity Press, p. 379.

12. Morris, P.W.G. (1994) *The Management of Projects,* London: Thomas Telford.
13. (n.d.) Life-cycle costing. http://dept.lamar.edu/industrial/Underdown/eng_mana/Life_Cycle_Costing_ch10.htm
14. (n.d.) Life-cycle costing. *Ibid.*
15. Dallas, M. F. (2006) *Value and Risk Management: A Guide to Best Practice.* Oxford, UK: Blackwell.
16. Wideman, M. (n.d.) Project value management. http://www.maxwideman.com/issacsons3/iac1338/sld001.htm)

KEY TERMS

Project value chain
Concurrent engineering
Kano model
Threshold attributes
Performance attributes
Excitement attributes
Interorganizational cost management

Target costing
Value engineering
Turnkey projects
Life-cycle costing (LCC)
Cost-estimating relationship (CER)

Index

ABC, *see* Activity-based costing
AC, *see* Actual cost of work performed
Accounting systems and practices, 84
Accreditation, 248
Action plans (VM), 9
Activities:
   accelerating (crashing) times for, 99–104
   in earned value analysis, 117
   forecasting time required for, *see* Forecasting methods
   impacting project values, 259
   repetitive, cost estimation for, 61–64
Activity-based costing (ABC), 90–93
   cost drivers in, 91
   examples of, 91–93
   steps in, 90–91
Activity cost estimating sheet, 58, 59
Actual cost of work performed (AC), 115, 118, 119, 121
Advanced development phase (LCC), 271
Airbus, 225
Airbus A380 project, 131
Alternative solutions, 22–23
AMEC Inc., 54–55
American National Standards Institute (ANSI), 192
ANSI/EIA-649– 1998, 192
Armored personnel carriers (APCs), 189–190
Association, linear regression technique for determining, 73
Assumptions:
   in cost estimation, 61
   scope document list of, 30

Audits:
   configuration, 195
   quality, 249
Auto catalytic growth function, 77

BAC (budgeted cost at completion), 115
BAE Systems, 215
Ballpark cost estimates, 51, 52
Baselines:
   budget, 93, 94
   configuration, 194
   project, 106, 116–117
   schedule, 93, 94
Benchmarking, 248, 249
Berkeley Project Management Process Maturity Model, 249
Big Dig, *see* Boston Central Artery/Tunnel project
Billing process, 147
Boeing Corporation, 52–53, 188, 215, 222, 225, 232
Boisvert, Henry, 190
Boston Central Artery/Tunnel (CA/T) project, 3–5
Bottom-up approach:
   in project budgeting, 88–90
   to work breakdown structure, 37
Bradley Fighting Vehicle, 189–190
Breakeven analysis, 131
Budgets:
   allocation of contingency funds to, 95–96
   contingency, 95–98
   fiscal operating vs. project, 83
   time-phased, 93–94, 114, 118
   in value and risk management integration, 273

Budgeted cost at completion (BAC), 115
Budgeting, *see* Project budgeting
Build–own–operate contracts, 154
Build–own–operate–transfer contracts, 154
Bullwhip effect, 211

Campbell Soup, 212
Capital, cost of, 154
Capital asset pricing model (CAPM), 153
Capital development projects, 128–129
CAPM (capital asset pricing model), 153
Cash float, 129, 130
Cash flow:
  applying discount rate to, 134–137
  concept of, 127–131
  defined, 127, 131
  and resource allocation over time, 84
Cash flow breakeven analysis, 131
Cash flow management, 127–148
  and concept of cash flow, 127–131
  defined, 127
  payment arrangements, 137–148
  and worth of projects, 131–137
CA/T project, *see* Boston Central Artery/Tunnel project
Causation, linear regression technique for determining, 73
Ceiling price, 145
Centre for Research on the Management of Projects (CRMP), 193
CER, *see* Cost estimating relationship
Certification, 247–248
Change(s):
  anticipating potential sources of, 196–197
  causes of, 186–190
  defined, 185
  influence of, 190–191
  positive/negative aspects of, 185
  of scope, 30, 32, 238–239
  to scope definition document, 29
  in specifications, 47–48, 239
Change control, 196–200
  potential sources of change, 196–197
  procedure for, 197–200
  responsibility for, 200
  for scope changes, 32
Change order processing time, 147
Channel Tunnel (Chunnel), England and France, 46, 128, 142, 152
Chinnagiri, Mahesh, 251
Chunnel, *see* Channel Tunnel, England and France
Claims, cost, 142–143
Clarity of purpose, 272
CM, *see* Configuration management
Coding, WBS, 38, 39
Coefficient of determination (R^2), 73, 74
Commercial risks, 159
Communication:
  in value and risk management integration, 273
  in value management, 9
Comparative cost estimates, 52–55
Completion deadlines, 273
Completion values, 124
Computerized project planning systems, 56–57
Conceptual design phase (LCC), 271
Conceptual development, 22–23
Concession agreements, 155
Concorde, 53–54
Concurrent engineering, 262
Configuration control system, 194, 197
Configuration management (CM), 191–195
  defined, 192
  example of, 202–206
  process for, 193–195
  standards for, 192
Constant rate assumption, 61
Constraints, identifying, 22
Contingencies, in cost estimation, 59–61, 95
Contingency budget, 95–98
Contract change notes, 143
Contractual risks, 159
Conversion, in project supply chain process, 224–225
Corporate culture, 245, 272–273
Corporate strategy, 17–18
Correlation coefficient (r), 74

Index

Cost(s):
    impact of needs on, 20
    predicting, *see* Forecasting methods
    as project performance metric, 228, 229
    relationship of time and, 45
    relationship of value and, 3–5
    sources and categories of, 49–50
    as value driver, 218
Cost control, 105–125
    earned value management, 111–125
    and project evaluation and control, 105–107
    time-cost analysis, 107–111
    time-phased budgets for, 93–94
Cost drivers (ABC), 91
Cost estimating relationship (CER), 268–269
Cost estimation, 43–64
    allowances for contingencies in, 59–61
    with computerized systems, 56–57
    contingency budget to offset errors in, 95
    detailed estimates, 56–60
    external factors affecting, 48
    importance of, 44–45
    keys for effectiveness in, 6–7
    "laws" of, 51
    learning curves in, 61–64
    low initial estimates, 46
    manual system for, 57–59
    methods for, 51–55
    process of, 56
    and quality of scope definition, 47
    for repetitive activities, 61–64
    sources/categories of project costs, 49–50
    and specification changes, 47–48
    and unanticipated technical difficulties, 46–47
    WBS as basis for, 27
Cost management, 2. *See also specific areas of management, e.g.:* Project budgeting
    effectiveness in, 8
    forecasting methods for, *see* Forecasting methods
    issues in, 3

Cost of capital, 154
Cost of debt, 154
Cost of equity, 153
Cost overruns, 45–49, 97–98
Cost performance index (CPI), 115, 121
Cost-plus-fee contracts, 138
Cost-plus reimbursement, 129, 138
Cost-reimbursable arrangements, 138–140
Cost variance (CV), 115, 118, 121
Cost variation, due to inflation and exchange rate fluctuation, 144–145
Counter trade, 153
CPI, *see* Cost performance index
Crash costs, 50
Crashing, *see* Project acceleration
Credit control system, 130
Crisis management, 201–202
Critical success factors (CSFs), 172
CRMP (Centre for Research on the Management of Projects), 193
Cross-functional framework (VM), 172
CSFs (critical success factors), 172
Customers:
    as critical SCM area, 213
    value chains unique to, 258
    as value drivers, 217–218
Customer satisfaction, Kano model of, 263–265
CV, *see* Cost variance

Data gathering, for budget preparation, 85
DCF (discounted cash flow), 132
Debt, cost of, 154
Debt/equity swapping, 153
Debt financing, 151–152
Decision making:
    for crashing project, 100–103
    cross-functional framework for, 172
    structured, 172–173
Definitive cost estimates, 52
Delays, changes causing, 198. *See also* Change control
Deliverables:
    cost estimates for, 51
    scope document list of, 29
    tying WBS elements to, 41

Index

Delivery:
   in project supply chain process, 225
   in SCOR model, 231
   in value and risk management integration, 273
Dell Computer Corporation, 218
Demand, forecasting, *see* Forecasting methods
Deming Prize, 248
Design:
   as critical SCM area, 214
   integrating cost and value in, 260–265
   requirements for, in statement of work, 24
Design modeling, 274
Design of experiments (DOE), 244
Detailed design phase (LCC), 271
Detailed evaluation phase, 173
Direct costs, 50, 102
Discounted cash flow (DCF), 132
Discount rate, 132, 134–137
Divestment phase (LCC), 271
Documents, SOW list of, 25
DOE (design of experiments), 244
Dulhasti Power project, India, 48

EAC (estimated cost at completion), 115
Earned value (EV), 112, 115, 121
Earned value analysis (EVA), 107
   conducting, 117–119
   elements in, 114
Earned value assessment, 119–122
Earned value management (EVM), 111–125
   conducting earned value analysis, 117–119
   effective use of, 123–125
   elements in earned value analysis, 114
   managing portfolio of projects with, 122–123
   model for, 112–113
   performing earned value assessment, 119–122
   relevancy of, 115–117
   terminology related to, 114–115
   time-phased budgets in, 93

Economic value added, 166
EMV (expected monetary value), 166–167
Equipment, cost of, 50
Equity, cost of, 153
Equity financing, 151
Error sum of squares (SSE), 74
Estimated cost at completion (EAC), 115
ETC, *see* Expected time to completion
European Quality Award, 248
European Union, 188
Eurotunnel, 142, 152
EV, *see* Earned value
EVA, *see* Earned value analysis
EVM, *see* Earned value management
Exchange rate fluctuation, 144–145
Excitement attributes (Kano model), 264
Expected monetary value (EMV), 166–167
Expected time to completion (ETC), 115, 121
Expedited costs, 50
Explained variation (linear regression), 74
Exponential growth or decay (in time series analysis), 69
Exxon/Mobil, 18

FAC (forecasted costs at completion), 115
Facilities, cost of, 50
Facility operation/maintenance (LCC), 271
Feasibility estimates, 52
Feasibility studies, 156
Feedback and control phase, 173
50/50 rule, 124
Financial management, 156–161
Financial package:
   arranging, 167
   controlling, 167–158
Financial risk, controlling, 158–159
Financing projects, 150–154
   cost of, 153–154
   defined, 150
   methods for, 127–128
   principles of, 150–151

Index

project finance vs., 149–150
  sources of finance, 153
  types of finance, 151–152
Finite element analysis, 274
Fiscal operating budgets, 83
Fixed costs, 50
Fixed-price contracts, 140–141
Flexibility:
  as project performance metric, 228, 229
  as supply chain value driver, 218
Float, 129, 130
FMC Corporation, 189, 190
Forecasted costs at completion (FAC), 115
Forecasting methods, 67–81
  categories of, 67–68
  linear regression analysis, 69–76
  S-curve, 77–81
  time series analysis, 69
Forfeiting, 153
Freezing of projects, 198
Funding projects, *see* Financing projects

Gap analysis, 107
General Electric Company, 167, 224
German Project Management Association, 248
Global Accreditation Center for Project Management, 248
Global Project and Procurement Network, 217
Goals, statement of, 22

The Handbook of Project-based Management, 219
Hewlett-Packard, 212
Hoechst Pharmaceuticals, 18
Human factor, in earned value management, 125

ICE (Institution of Civil Engineers), 168
Idea generation, 173
Inbound supply chain, integrating cost and value in, 260

Indirect costs, 50, 102
Individual events, forecasting, 67, 68
Inflation, cost variation due to, 144–145
Information gathering, 22
Information technology (IT) projects, 2, 44
Institution of Civil Engineers (ICE), 168
Integrating cost and value, 255
  in inbound supply chain, 260
  and integrated value and risk management, 272–274
  process for, 274–277
  in project delivery/implementation, 267–272
  in project design, 260–265
  in project development, 265–267
  strategies for, 261
Internal rate of return (IRR), 133–137, 159
International Organization for Standardization (ISO), 223
International Project Management Award, 248
Internet Week, 215
Interorganizational cost management, 266
Inventory, as critical SCM area, 215
IRR, *see* Internal rate of return
ISO (International Organization for Standardization), 223
ISO-9000/9001, 223
IT projects, *see* Information technology projects

Joint Fight Striker (JFS) program, 29, 215–217, 226
Journal of Business Strategy, 209
Jowell, Tessa, 2

Kaminsky, Phil, 213
Kano model, 263–265
Key performance indicators (KPIs), 172
Kodak, 242
KPIs (key performance indicators), 172

Labor cost, 49
  under constant rate assumption, 61
  and learning curve, 61–64
LCC, *see* Life-cycle costing
Leadership, 272–273
Lean manufacturing, 265–267
Lean production, 212
Learning curves, in cost estimation, 61–64
Leasing assets, 153
Least squares regression, 71
Lessons learned, 202
Life cycle, project, *see* Project life cycle
Life-cycle costing (LCC), 268–272
Limit-of-liability contracts, 138
Linear regression analysis, 69–76
  best fit in, 70–71
  evaluating fit of regression line, 73–76
  example of, 71–73
  interpretation of coefficients, 71–73
  least squares regression, 71
  linear trend (straight line) equation in, 70
  multiple linear regression, 76
Linear trend (time series analysis), 69
Lockheed Martin, 2, 215
Logistics, as critical SCM area, 214
London Olympics, 2
Lump sum contracts, 140–141

"Make" process (SCOR model), 231
Malcolm Baldridge Award, 248
The Management of Business Logistics, 230
"Management of projects," 268, 276
Materials, cost of, 49
Metz, Peter J., 212
Mezzanine debt, 152
Miles, Lawrence D., 168
Milestones, identifying, 93
Millennium Bridge, London, 47
Millennium Dome, London, 43–44
MIL-STD-973, 192
Money, time value of, 132–133
Morris, Peter, 268
Motorola Corporation, 249
Multiple linear regression, 76
Multiyear projects, budgeting for, 84

"Must-be" attributes (Kano model), 263, 264

National Bicycle, 212
Needs assessment, 8, 19–22, 173
Net present value (NPV), 134–137, 159–161, 166–167
Nonlinear trend (in time series analysis), 69
Nonrecurring costs, 50
Normal costs, 50
Northrop Grumman, 215
No trend (in time series analysis), 69
NPV, *see* Net present value

Objectives:
  project, 23
  in scope definition document, 29
OBS, *see* Organizational breakdown structure
OD (original duration), 115
Off-take contracts, 155
"One-dimensional" attributes (Kano model), 263, 264
Operations, as critical SCM area, 214
Optimum choice phase, 173
Options models, 159–161
Organizational breakdown structure (OBS), 34, 35
Original duration (OD), 115
Outcome of project, forecasting, 68. *See also* Forecasting methods

Parametric cost estimation, 52–55
Payment arrangements:
  and cash flow, 137–148
  claims and variations, 142–143
  cost-plus, 129
  cost-reimbursable, 138–140
  cost variation due to inflation/exchange rate fluctuation, 144–145
  credit control system for, 130
  payment plans, 140–142
  price incentives, 145–146
  retentions, 146–147
Payment plans, 140–142

Index

Pay scales, 147
Pearl curve, *see* S-curve
Pentagon, JFS program and, 215, 216
Percentage complete rule, 124–125
Performance attributes (Kano model), 263, 264
Performance monitoring and tracking, 19
   gap analysis for, 107
   in project control process, 106–107
   of project supply chain, 227–229
   statement of work in, 24
   time–cost analysis for, 107–111
   in value management, 9
Performance variations, price adjustments due to, 146
Pfizer, 18
PING Golf, Inc., 273–274
Planned value (PV), 114, 118, 121
Planning, *see* Project planning
Planning phase (SCOR model), 231
PMBoK (Project Management Institute), 33, 191
PMI, *see* Project Management Institute
Porter, Michael, 255
Portfolio of projects, earned value management with, 122–123
Pratt & Whitney, 215
Price incentives, 145–146
Problem definition/statement, 23
   clarity of, 19
   by project team, 23
Process-oriented WBS, 34, 36
Procter & Gamble, 209
Procurement:
   in project supply chain process, 221–224
   in value and risk management integration, 273
Product-based WBS, 33–34, 38, 39
Production phase (LCC), 271
Product quality management, 235–236. *See also* Quality management
Product standards, 252
Program budgeting, 93–94
Progress monitoring:
   forecasting in, 67–68. *See also* Forecasting methods
   in project control process, 106–107
Project acceleration ("crashing"), 99–104
   costs associated with, 45, 50
   decision making for, 100–103
   and project budgeting, 99–104
   repeating process of, 101
   use of contingency funds for, 98
Project-based ventures, 1, 18
Project baseline plan, 106
Project budgeting, 83–104
   activity-based costing, 90–93
   approaches to, 85–90
   bottom-up, 88–90
   contingency budget in, 95–98
   cost estimation linked to, 45
   developing budget, 85
   issues in, 83–84, 98
   and program budgeting, 93–94
   and project acceleration ("crashing"), 99–104
   top-down, 86–88
Project closeout management, 147
Project control process, 106–107
Project delivery/implementation, integrating cost and value in, 267–272
Project development, integrating cost and value in, 265–267
Project evaluation and control system, 105–107
Project execution plan, 176
Project finance, 154–156. *See also* Financial management
   contingencies in, 59
   defined, 150
   financing of projects vs., 149–150
   planning, 156–157
Project life cycle, 18–19. *See also specific stages*, e.g.: Needs assessment
   accuracy of cost estimates through, 48–49
   life-cycle costing, 268–272

Project Management Institute (PMI), 33, 191, 248
Project management standards, 252
Project objectives, 23
Project planning, 27–41
　creation of work breakdown structure vs., 41
　as part of project strategy, 17
　scope definition, 28–32
　in value management, 8
　work breakdown structure in, 32–41
Project quality management, 236. *See also* Quality management
Project standards, 252
Project strategy, 17
Project supply chain process framework, 221–225
Project team:
　contingency budget input from, 97, 98
　problem definition by, 23
Project value, 8, 163. *See also* Value management
　evaluating, 17
　fundamental concepts of, 164–166
　risk management and increase in, 181
　tracking, 9
Project value chain, 256–257
Project value chain analysis, 257–259
Proof-of-concept, 180–181
Proposals:
　development of, in value management, 9
　standards for evaluating, 24
PV, *see* Planned value

QA (quality assurance), 244
Qualitative methods:
　for forecasting, 68
　for progress/performance measurement, 106
Quality:
　as project performance metric, 228, 229
　as supply chain value driver, 219
Quality assurance (QA), 244
Quality award models, 248

Quality Champion, 245
Quality control, 245
Quality engineering, 239–243
Quality loss function, 240
Quality management, 235–253
　methods for project organizations, 247–252
　process of, 236–237
　for product quality, 235–236
　for project quality, 236–245
　and quality standards for projects, 252–253
　Six Sigma methodology for, 249–252
　total quality management for, 245–247
Quantitative methods:
　for forecasting, 68
　for progress/performance measurement, 106, 107

r (correlation coefficient), 74
R^2, *see* Coefficient of determination
Rampey, J., 220
Recurring costs, 50
Regression analysis, 69. *See also* Linear regression analysis
Regression sum of squares (SSR), 74
Repetitive activities, learning curve for, 61–64
Requests for quotes (RFQs), 52
Resource allocation. *See also* Project budgeting
　methods used for, 85
　over time periods, 84
Retentions, 146–147
Return phase (SCOR model), 231
Reviews:
　quality, 249
　value management, 176–179
RFQs (requests for quotes), 52
Risk management, 180–184
　and crisis control, 202
　financial risk control, 158–159
　integrated value management and, 272–274
　risk assessment, 180–182
　and value management, 181–184
　WBS level of detail for, 41
Risk register, 182

Roberts, Harry V., 220
Robust design, 240–243
Rolls-Royce, 215, 222
The Royal Bank of Canada, 18

SCC (Supply Chain Council), 230
Schedules:
   for activities and resource usage, 118
   freezing, 199
   project, 27, 93, 199
Schedule performance index (SPI), 115, 121
Schedule variance (SV), 115, 118
Schedule variations, price adjustments due to, 146
SCM, see Supply chain management
Scope:
   changes in, 30, 32, 47–48, 238–239
   freezing, 199
Scope analysis (VM), 8
Scope creep, 47–48
Scope definition, 28–31
   common shortcomings in, 29–30
   cost estimates and quality of, 47
   elements of, 28–30
   for house construction project, 31
   list of assumptions in, 30
   outcome of, 27
   in project planning, 28–32
   and statement of work, 28
Scope statement:
   in statement of work, 25
   statement of work vs., 27
SCOR model, 230–231
S-curve:
   in forecasting, 77–81
   in performance monitoring, 109–111
   significant shortcomings of, 111
Senior debt, 151–152
Shui On Construction Company, 220–221
Sikorsky Aircraft, 222–223
Simchi-Levi, David, 213
Simchi-Levi, Edith, 213
Six Sigma, 249–252
   model for projects, 250–251
   software PM application of, 251–252
Solheim, John K., 274

Sourcing phase (SCOR model), 231
SOW, see Statement of work
Special purpose or project vehicle (SPV) stakeholders, 154
Specifications:
   changes in, 47–48, 239
   freezing, 198
SPI, see Schedule performance index
Sport Obermeyer, 212
SPV (special purpose or project vehicle) stakeholders, 154
SSE (error sum of squares), 74
SSR (regression sum of squares), 74
SST, see Total sum of squares
Stage payments, 141–142
Stakeholders:
   defining needs of, 19
   perceptions of value among, 3
   special purpose/project vehicle, 154
Standards:
   for configuration management, 192
   for proposal evaluation, 24
   quality, 252–253
   for work, 24
Standard error of the estimate (s_{yx}), 74–75
Statement of goals, 22
Statement of work (SOW), 24–26
   form for, 25–26
   project scope vs., 27
   scope definition document vs., 28
   topics included in, 24–25
Status accounting, configuration, 194
Strategies, corporate vs. project, 17
Structured decision making, 172–173
Submittal process, 147
Suppliers:
   as critical SCM area, 213–214
   development of, 224
Supplier Kaizen, 223, 224, 260
Supply chains:
   choosing, 221
   components of, 210–211
   integration of, 225–227
   optimizing value in, 220–221

project supply chain process
  framework, 221–225
total integration of, 225–227
Supply Chain Council (SCC), 230
Supply chain management (SCM),
  209–232
  benefits of, 212–213
  critical areas of, 213–215
  defined, 210
  future issues in, 231–232
  need for, 211–212
  to optimize value in supply chains,
    220–221
  performance metrics in, 227–229
  in project management, 215–217
  and project supply chain process
    framework, 221–225
  and SCOR model, 230–231
  for total integration of supply chain,
    225–227
  value drivers in, 217–219
Supply chain operations reference
  model, see SCOR model
SV, see Schedule variance
Switch trades, 153
s_{yx}, see Standard error of the estimate

Taguchi, Genichi, 239–243
Target costing, 266–267
Target price, 145
Task budgets. See also Activity-based
  costing
  bottom-up approach for, 88
  contingency fund allocation to,
    95–96
  top-down approach for, 86–88
Technical specifications, 28
Technology forecasting, 68. See also
  Forecasting methods
3M, 18
Threshold attributes (Kano model),
  263, 264
Time:
  accelerating (crashing), 99–103
  for activities, forecasting, see
    Forecasting methods
  as project performance metric,
    227–229
  relationship of cost and, 45

resource allocation over, 84
  as supply chain value driver, 219
Time-cost analysis, 107–111
Time–cost trade-off curve, 99
Time-phased budgets, 93–94, 114, 118
Time series analysis, 69
Time value of money, 132–133
Top-down approach:
  in project budgeting, 86–88
  to work breakdown structure, 37
Total quality management (TQM),
  212, 219–221, 245–247
Total sum of squares (SST), 73–74
TQM, see Total quality management
Trade-off analysis (VM), 9
Transmanche Link, 142
Trend (in time series analysis), 69
Turner, Rodney, 6
Turnkey projects, 268

Uncertainty:
  and complexity of supply chain, 212
  contingency budget for, 95
  in cost estimation, 45, 46, 48
  in life-cycle costing, 269–270
  as risk, 180–181
Unexplained variation (linear
  regression), 74
University of Manchester, 193

VA, see Value analysis
VAC (variance at completion), 115
Value. See also Project value
  adding, 164
  defined, 3, 163
  dimensions and measures of,
    166–167
  impact of needs on, 20
  as multidimensional concept, 3
  relationship of cost and, 3–5
Value analysis (VA), 170–171
Value chains, 210, 254. See also
  Supply chains
Value drivers:
  in supply chain management,
    217–219
  in value chain analysis, 258

Value engineering (VE), 170, 177, 261–262, 266
Value management (VM), 2, 167–184. *See also* Earned value management
  benefits of, 175
  defined, 167–168
  emphasis in, 3
  forecasting methods for, *see* Forecasting methods
  fundamental concepts in, 164–166
  goal of, 163–164
  integrated risk management and, 272–274
  key attributes of, 169
  key concepts in, 171–173
  key features of, 8–9
  key principles of, 168
  need for, 171
  process for, 173–174
  reviews of, 176–179
  and risk management, 180–184
  scope of, 168
  stages in, 176
  terminology related to, 169–171
Value planning (VP), 170
Variable costs, 50
Variance at completion (VAC), 115
Variations, payment, 142, 143
Variation orders, 143
VE, *see* Value engineering
Venezuela, oil development nationalization in, 158
VM, *see* Value management
VP (value planning), 170

Wal-Mart, 212–213
WBS, *see* Work breakdown structure
Weyerhaueuser Corporation, 18
Work breakdown structure (WBS), 32–41
  as basis for budgeting, 84, 88
  as basis for schedules/cost estimates, 27
  bottom-up approach to, 37
  coding of, 38, 39
  development of, 35–38
  in earned value analysis, 114
  function of, 32–33
  guidelines for developing, 41
  hierarchy of, 33
  integrating organization and, 38, 40
  levels in, 36–37, 41
  in project baseline plan, 106
  purpose of, 27
  top-down approach to, 37
  types of, 33–35
Work packages, 33
  contingency fund allocation for, 96, 97
  cost estimates based on, 51
  earned value calculation for, 116
  time-phased budgets for, 114
  top-down budgeting for, 86
  in WBS development, 36
Work scope document, 27–30. *See also* Scope definition
Worth of projects, 131–137

0/100 rule, 124

Notes

Notes

Notes

Notes

Notes

Notes

Notes

Notes

Notes

Notes

Notes

Notes

Notes